ASTRONOMICAL NUMBERS

By Thomas Wm. Hamilton

Strategic Book Publishing and Rights Co.

Copyright © 2016 Thomas Wm. Hamilton. All rights reserved.

No part of this book may be reproduced or transmitted in any form or by any means, graphic, electronic, or mechanical, including photocopying, recording, taping, or by any information storage retrieval system, without the permission, in writing, of the publisher. For more information, send an email to support@sbpra.net, Attention Subsidiary Rights Department.

Strategic Book Publishing and Rights Co., LLC
USA | Singapore

For information about special discounts for bulk purchases, please contact Strategic Book Publishing and Rights Co. Special Sales, at bookorder@sbpra.net.

ISBN: 978-1-68181-490-2

Book Design: Suzanne Kelly

INTRODUCTION

Though much of the information found in this book is available on the Internet, such a search could take hours. I have presented it here for the convenience of the reader, and in a format easily understood by both the novice and the astronomical specialist. Data covered herein includes the Sun, the dozen nearest stars, the dozen brightest stars, the nearest Brown Dwarfs, the dozen nearest known exoplanets, our Local Group of Galaxies, all the planets, those asteroids that have been visited by spacecraft, and selected moons, including all that have been visited by spacecraft, plus the largest craters and tallest mountains on all of these. Also included are the various distance units used by astronomers: kilometers, miles, Astronomical Units, lightyears, and parsecs, and their mutual relationships.

The list of nearest stars is probably complete. The same cannot be said for nearest Brown Dwarfs or exoplanets. A mere glance at the discovery dates should prove these are still being found.

To make this work as useful as possible to the widest audience, nearly all values are given in both metric and English units.

Moons that have not been examined by spacecraft, as well as those that have been, are all treated in an earlier book, *Moons of the Solar System*. A list of all my books that are currently in print may be found at the end of this book.

Adrastea

(second moon outward from Jupiter, third moon discovered to orbit in less than its planet's day)

Albedo: 0.80

Apparent Magnitude: +18.7 (at Jupiter's mean opposition)

Axial Tilt (Obliquity): 0

Density: ~0.9

Diameter: 20 by 16 by 14 kilometers; 12.5 by 10 by 8.5 miles

Discovery: July 9, 1979 by David Jewett

Escape Velocity: 28 feet/second; 8 meters/second

Mass: 2×10^{15} kilograms

Orbit
    Apoapsis: 133,100 kilometers; 81,900 miles
    Periapsis: 126,900 kilometers; 79,300 miles
    Semi-major axis: 129,000 kilometers; 80,100 miles
    Eccentricity: 0.0015
    Inclination to planet's equator: 0.03 degree
    Period: 7 hours 9.5 minutes; 0.29826 Earth day
    Average speed: 31.378 kilometers/second; 19.5 miles/second

Rotation: Synchronous with orbit

Surface Gravity: 0.002 meters/second^2; 0.00022g (1g=Earth's gravity on the surface of Earth)

Spacecraft
    Voyager 1...........1979
    Voyager 2...........1979
    Galileo................2002

Amalthea

(third moon outward from Jupiter; Jupiter V)

Albedo: 0.09

Apparent Magnitude: +14.1 (at Jupiter's mean opposition)

Axial Tilt: 0

Density: 0.857 (water = 1)

Diameter: 250 by 146 by 128 kilometers; 155 by 90 by 80 miles

Discovered: September 9, 1892 by Edward Emerson Barnard. Last moon discovered by eyesight/telescope rather than photography.

Escape Velocity: 0.058 kilometer/second; 0.036 mile/second

Mass: 2.08×10^{18} kilograms

Orbit
    Apoapsis: 182,840 kilometers; 113,611 miles
    Periapsis: 181,150 kilometers; 112,561 miles
    Semi-major axis: 181,365.84 kilometers; 2.54 Jovian radii; 112,590 miles
    Eccentricity: 0.00319
    Inclination to planet's equator: 0.374 degree
    Period: 0.49817943 Earth day; 11 hours 57 minutes 23 seconds
    Average speed: 26.57 kilometers/second; 16.51 miles/second

Rotation: Synchronous with orbit

Surface Gravity: 0.020 meters/second^2, 0.0022g

Spacecraft
    Voyager 1...............1979
    Voyager 2...............1979
    Galileo....................2002

Ariel

(Uranus I; fifteenth moon outward from Uranus; fourth largest of Uranus' moons)

Albedo: 0.53

Apparent Magnitude: +14.4 at planet's mean opposition

Axial Tilt: 0

Density: 1.592 (water = 1)

Diameter: 1162.2 by 1155.8 by 1155.4 kilometers; 722 by 718 by 718 miles

Discovery: October 24, 1851 by William Lassell

Escape Velocity: 0.559 kilometer/second; 0.347 mile/second

Mass: 1.353 X 10^{21} kilograms

Orbit

    Apoapsis: 191,200 kilometers; 118,806 miles

    Periapsis: 190,800 kilometers; 118,558 miles

    Semi-major Axis: 191,020 kilometers; 118,623 miles

    Eccentricity: 0.0012

    Inclination to planet's equator: 0.260 degree

    Period: 2.520 Earth Days

    Average Orbital Speed: 5.51 kilometers/second; 3.5 miles/second

Rotation: synchronous to orbit

Surface Gravity: 0.269 meters/second^2; 0.029g

Spacecraft

Voyager 2..............January 26, 1986..........fly-by

Ariel on January 24, 1986 from 130,000 kilometers/80,000 miles.

ASTRONOMICAL UNIT

The Astronomical Unit (abbreviated A.U.) is basically Earth's average distance from the Sun. Earth has an elliptical orbit (see *Earth: Orbit* -- Aphelion or Perihelion or Eccentricity). Measurements of this figure and its formal definition have varied over the centuries. The value currently used for the Astronomical Unit is:

149,957,870 kilometers

92,955,807 miles

Light travels one A.U. in 499.004783806 seconds (8 minutes = 480 seconds, so 1 A.U. in 8 minutes 19 seconds is a fair approximation).

0.000015812507 lightyear = 1 A.U.

0.00000481368 parsec = 1 A.U.

Brown Dwarfs

Nearest Earth (all those known within twelve lightyears)

Luhman 16A
    Apparent Magnitude: +10.7
    Absolute Magnitude: +14.2
    Constellation: Vela (10h49m18.7s; -53 degrees 19' 09.9")
    Discovery: 2010 by Kevin Luhman
    Distance: 6.516 Lightyears
    Mass: 0.04 (Sun = 1)
    Spectral Type: L7.5

Luhman 16B
    Apparent Magnitude: +11
    Absolute Magnitude: +15
    Constellation: Vela (10h49m18.7s; -53 degrees 19' 09.9")
    Discovery: 2010 by Kevin Luhman
    Distance: 6.516 lightyears
    Mass: 0.03 (Sun = 1)
    Spectral Type: T0.5

WISE 0855-0714
    Apparent Magnitude: +13.89
    Absolute Magnitude: +17.2
    Constellation: Hydra (08h55m10.8s; -07 degrees 14' 42.5")
    Discovery: March 2013 by Kevin Luhman
    Distance: 7.53 lightyears
    Mass: <0.08 (Sun = 1)
    Spectral Type: Y2

WISE 1506+7027
    Apparent Magnitude: +14.3
    Absolute Magnitude: +16.6
    Constellation: Ursa Minor (15h06m; +70 degrees 27')
    Discovery: 2011 by J. D. Kirkpatrick
    Distance: 11.09 lightyears
    Mass: <0.08 (Sun = 1)
    Spectral Type: T6

Callisto

(eighth moon outward from Jupiter; second largest of Jupiter's moons; Jupiter IV)

Albedo: 0.22

Apparent Magnitude: +5.65 at planet's mean opposition

Atmosphere: CO_2; pressure 7.5×10^{-12} bar

Axis Tilt: 1 degree

Density: 1.83 (water = 1)

Diameter: 4821 kilometers; 2995 miles; 0.378 Earth

Discovery: January 8, 1610 by Galileo Galilei

Escape Velocity: 2.44 kilometers/second; 1.52 miles/second

Mass: 1.8344×10^{23} kilograms

Orbit
    Apoapsis: 1,897,000 kilometers; 1,178,800 miles
    Periapsis: 1,869,000 kilometers; 1,161,400 miles
    Semi-Major Axis: 1,882,700 kilometers; 1,169,700 miles
    Eccentricity: 0.0074
    Inclination: 0.192 degree to planet's equator
    Period: 16.6890184 Earth days; 16 days 16 hours 32.2 minutes
    Average Orbital Speed: 8.2 kilometers/second; 5.1 miles/second

Rotation: Synchronous with orbit

Spacecraft
    Voyager 1............1979
    Voyager 2............1979
    Galileo................2002

Callisto from the Galileo orbiter in June 2005.

Ceres

(First asteroid or dwarf planet discovered, largest)

Albedo: 0.09

Apparent Magnitude: +6.64 to +9.34

Density: 2.16

Diameter: 965.2 by 961.2 by 891.2 kilometers; 600 by 597 by 553.8 miles

Discovered: January 1, 1801 by Giuseppe Piazzi

Escape Velocity: 0.51 kilometer/second; 0.315 mile/second

Mass: 9.393×10^{20} kilograms

Orbit
    Aphelion: 445,410,000 kilometers; 2.9773 Astronomical Units; 276,765,000 miles
    Perihelion: 382,620,000 kilometers; 2.5577 Astronomical Units; 237,749,000 miles
    Semi-Major Axis: 414,010,000 kilometers; 2.7670322 Astronomical Units; 257,254,000 miles
    Eccentricity: 0.075823
    Inclination to ecliptic: 10.593 degrees
    Period: 4.60 years; 1681.601 Earth days
    Average Orbital speed: 17.909 kilometers/second; 11.128 miles/second

Rotation: 0.3781 day; 9.07417 hours; 9 hours 4 minutes

Surface Gravity: 0.29 meter/second^2; 0.03g

Spacecraft
    Dawn..............March 6, 2015........orbiter

Ceres from 385 kilometers/240 miles showing Occator Crater.

Chariklo

(asteroid; first asteroid found to have rings)

Albedo: 0.045

Apparent Magnitude: +18.3 at mean opposition

Density: unknown

Diameter: 302 kilometers;

Discovery: February 15, 1997 at Spacewatch by James V. Scotti & Felipe Braga-Ribas

Escape Velocity: unknown

Mass: unknown

Orbit
    Aphelion: 2,766,857,500 kilometers; 1,719,245,500 miles; 18.4953 Astronomical Units
    Perihelion: 1,953,673,400 kilometers; 1,213,956,400 miles; 13.0595 Astronomical Units
    Semi-Major Axis: 2,360,265,000 kilometers; 1,466,601,000 miles; 15.7774 Astronomical Units
    Eccentricity: 0.172265
    Inclination: 23.4085 degrees
    Period: 62.67 years; 22,890.29 days

Rotation: 7.004 hours

Rings

A
    Width: 7 kilometers; 4 miles
    Distance: 396 kilometers; 246 miles

B
    Width: 3 kilometers; 1.9 miles
    Distance: 405 kilometers; 252 miles

Charon
(closest & largest moon of Pluto)

Albedo: 0.38

Apparent Magnitude: +16.8 to +17.3 at planet's opposition

Density: 1.720 (water = 1)

Discovered: June 22, 1978 by James Christy

Escape Velocity: 0.59 kilometer/second; 0.37 mile/second

Mass: 1.59 X 10^21 kilograms

Orbit
    Apoapsis: 19,600 kilometers; 12,160 miles
    Periapsis: 19,600 kilometers; 12,160 miles
    Semi-Major Axis: 19,600 kilometers; 12,160 miles
    Eccentricity: 0
    Inclination: 0
    Period: 6.38270 days; 6 days 9 hours 17 minutes 36 seconds

Rotation: Synchronous with orbit

Surface Gravity: 0.29 meter/second^2; 0.03g

Spacecraft
    New Horizons.............July 14, 2015...........fly-by planet

Charon during fly-by, 2015.

CONSTELLATIONS

The list below shows all 88 recognized constellations by *Name*, *Source* (A for Ancient, generally long before the First Century A.D., M for Modern, generally created between 1500 and 1800, and (A) for ones formed out of Argo); *Place* (E for ecliptic, Q for sitting athwart the celestial equator, N for north of the equator, and S for south of the equator), *Size* in square degrees on the celestial sphere, and *Rank* its size order.

Name	Source	Place	Size	Rank
Andromeda	A	N	722	19
Antlia	M	S	239	62
Apus	M	S	206	67
Aquarius	A	EQ	980	10
Aquila	A	Q	652	22
Ara	A	S	237	63
Aries	A	EN	441	39
Auriga	A	N	657	21
Bootes	A	N	907	13
Caelum	M	S	125	81
Camelopardalis	M	N	757	18
Cancer	A	EN	506	31
Canes Venatici	M	N	465	38
Canis Major	A	S	380	43
Canis Minor	A	N	183	71
Capricornus	A	ES	414	40
Carina	(A)	S	494	34
Cassiopeia	A	N	598	25
Centaurus	A	S	1060	10
Cepheus	A	N	588	27
Cetus	A	Q	1231	4

Chamaeleon	M	S	132	79
Circinus	M	S	93	85
Columba	M	S	270	54
Coma Berenices	A	N	386	42
Corona Australe	A	S	128	80
Corona Borealis	A	N	179	73
Corvus	A	S	184	70
Crater	A	S	282	53
Crux	M	S	68	86
Cygnus	A	N	804	16
Delphinus	A	N	189	69
Dorado	M	S	179	72
Draco	A	N	1083	8
Equuleus	A	N	72	87
Eridanus	A	S	1138	6
Fornax	M	S	398	41
Gemini	A	EN	514	30
Grus	M	S	366	45
Hercules	A	N	1225	5
Horologium	M	S	249	58
Hydra	A	Q	1303	1
Hydrus	M	S	243	61
Indus	M	S	294	49
Lacerta	A	N	201	68
Leo	A	EN	947	12
Leo Minor	M	N	232	64
Lepus	A	S	290	51
Libra	A	ES	538	29
Lupus	A	S	334	46
Lynx	M	N	545	28

Lyra	A	N	286	52
Mensa	M	S	153	75
Microscopium	M	S	210	66
Monoceros	M	Q	482	35
Musca	M	S	138	77
Norma	M	S	165	74
Octans	M	S	291	50
Ophiuchus	A	Q	948	11
Orion	A	Q	594	26
Pavo	M	S	378	44
Pegasus	A	N	1121	7
Perseus	A	N	615	24
Phoenix	M	S	469	37
Pictor	M	S	247	59
Pisces	A	EQ	889	14
Pisces Austrinus	A	S	245	60
Puppis	(A)	S	673	20
Pyxis	M	S	221	65
Reticulum	M	S	114	82
Sagitta	A	N	80	86
Sagittarius	A	ES	867	15
Scorpius	A	ES	497	33
Sculptor	M	S	475	36
Scutum	M	S	109	84
Serpens	A	Q	637	23
Taurus	A	EN	797	17
Telescopium	M	S	252	57
Triangulum	A	N	132	78
Triangulum Australe	M	S	110	83
Tucana	M	S	295	48

Ursa Major	A	N	1280	3
Ursa Minor	A	N	256	56
Vela	(A)	S	500	32
Virgo	A	EQ	1294	2
Volans	M	S	141	76
Vulpecula	M	N	268	55

CRATERS, LARGEST IMPACT

Object	Crater Name	Diameter
Mercury	Caloris Basin	1550 kilometers; 960 miles
Venus	Mead Crater	280 kilometers; 175 miles
Earth	Ross Crater (Antarctica)	540 kilometers; 350 miles*
	Vredevort (South Africa)	300 kilometers; 185 miles
Moon	Oceanus Procellarum	3000 kilometers; 2000 miles
Mars	Utopia Planitia	3300 kilometers; 2180 miles
Phobos	Stickney Crater	9 kilometers; 5.6 miles
Deimos	Voltaire Crater	1.9 kilometers; 1.2 miles
Eros	Charlois Regio	1.2 kilometers; 0.75 mile
Vesta	Rheasylvia Crater	505 kilometers; 310 miles
Ceres	Kerwen Crater	284 kilometers; 180 miles
Ida	Lascaux Crater	12 kilometers, 7.5 miles
Dactyl	Acmon Crater	300 meters; 950 feet
Lutetia	Massilia Crater	61 kilometers; 37.9 miles
Itokawa	Miyabaru Crater	0.609 kilometer; 0.378 mile
Steins	Diamond Crater	2.1 kilometers; 1.3 miles
Gaspra	Saratoga Crater	2.8 kilometers; 1.74 miles
Mathilde	Karoo Crater	33.4 kilometers; 21.6 miles
Amalthea	Pan Crater	89 kilometers; 55 miles

Thebe	Zanthus Crater	~22 kilometers; ~13 miles
Europa	Midir Crater	37.4 kilometers; 23.24 miles
Ganymede	Epigeus Crater	343 kilometers; 215 miles
Callisto	Valhalla Basin	3800 kilometers; 2500 miles; largest crater in Solar System
Janus	Lynceus Crater	22 kilometers; 12.3 miles
Mimas	Herschel Crater	139 kilometers; 86 miles
Tethys	Odysseus Crater	445 kilometers; 277 miles
Dione	Evander Crater	350 kilometers; 220 miles
Rhea	Mamaldi Crater	480 kilometers; 300 miles
Titan	Manrva Crater	392 kilometers; 244 miles
Enceladus	Aladdin Crater	37.4 kilometers; 23.1 miles
Hyperion	Jarilo Crater	121.57 kilometers; 75.54 miles
Iapetus	Turgis Crater	550 kilometers; 360 miles
Phoebe	Jason Crater	~60 kilometers; ~40 miles
Miranda	Alonso Crater	25.0 kilometers; 15.5 miles
Ariel	Yangoor Crater	78 kilometers; 48.5 miles
Umbriel	Wokolo Crater	210 kilometers; 130 miles
Titania	Gertrude Crater	326 kilometers; 203 miles
Oberon	Hamlet Crater	206 kilometers; 128 miles
Triton	Mazomba Crater	27 kilometers; 16.5 miles
Pluto	Burney Crater	200 kilometers; 124 miles
Charon	Mordor Crater	475 kilometers; 295 miles

* Disputed

Dactyl
(moon of asteroid Ida)

Albedo: ~0.2

Density: ~2.2

Diameter: 1.6 by 1.4 by 1.2 kilometers; 1 by 0.87 by 0.75 mile

Discovery: August 28, 1993 by Ann Horch

Orbit
    Semi-Major Axis: ~90 kilometers; ~56 miles
    Inclination to asteroid's equator: 8 degrees
    Period: 20 hours
    Average Orbital Speed: ~10 meters/second; ~33 feet/second

Spacecraft
    Galileo...........August 28, 1993........................fly-by

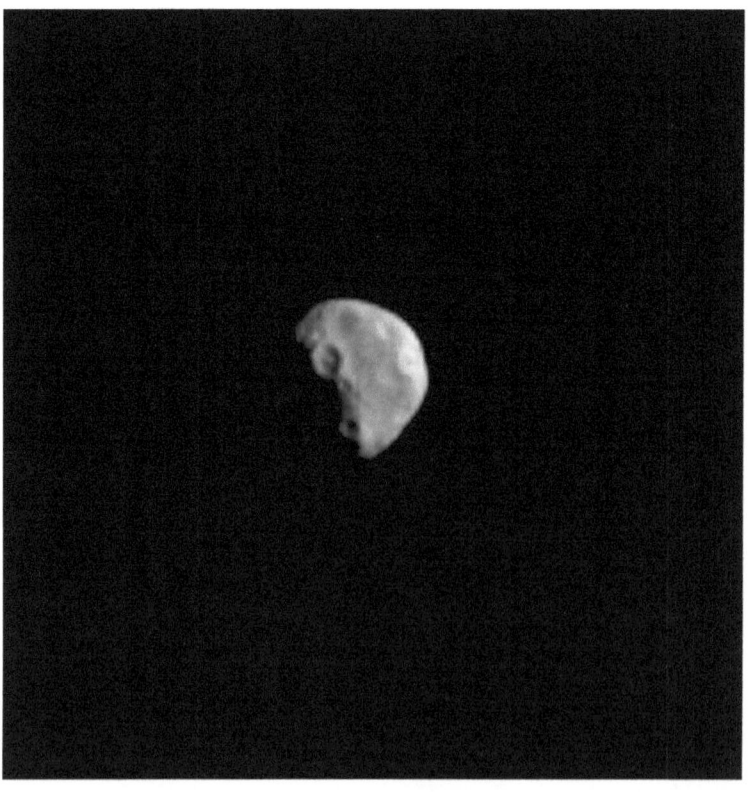

Asteroidal moon Dactyl from 3900 kilometers/2400 miles on June 4, 1994.

Deimos

(outer moon of Mars)

Albedo: 0.068

Apparent Magnitude: +12.45 at planet's mean opposition

Density: 1.471

Diameter: 15 by 11 by 10.3 kilometers; 9.3 by 7.6 by 6.4 miles

Discovery: August 1877 by Asaph & Angelina Stickney Hall

Escape Velocity: 5.556 meters/second; 20 kilometers/hour; 12.8 miles/hour

Mass: 1.4762×10^{15} kilograms

Orbit

Apoapsis: 23,470.9 kilometers; 14,584 miles

Periapsis: 23,455.5 kilometers; 14,574.6 miles

Semimajor axis: 23,463.2 kilometers; 14,580 miles

Eccentricity: 0.00033

Inclination to Mars' equator: 0.93 degree

Orbital Speed, average: 1.3513 kilometers/second; 0.84 mile/second

Period: 1.26244 Earth days; 30.312 hours

Synodic period: 1.2643 Martian days

Rotation: Synchronous with orbit

Surface Gravity: 0.0003 meters/second; 0.000032g

Spacecraft Photography

Mariner 9	1971	from space
Viking 1	1977	from space
Global Surveyor	1998	from space
Mars Express	2004	from space
Opportunity	2004	from Mars' surface
Spirit	2004	from Mars' surface
Reconnaissance	2007	from space
Curiosity	2013	from Mars' surface

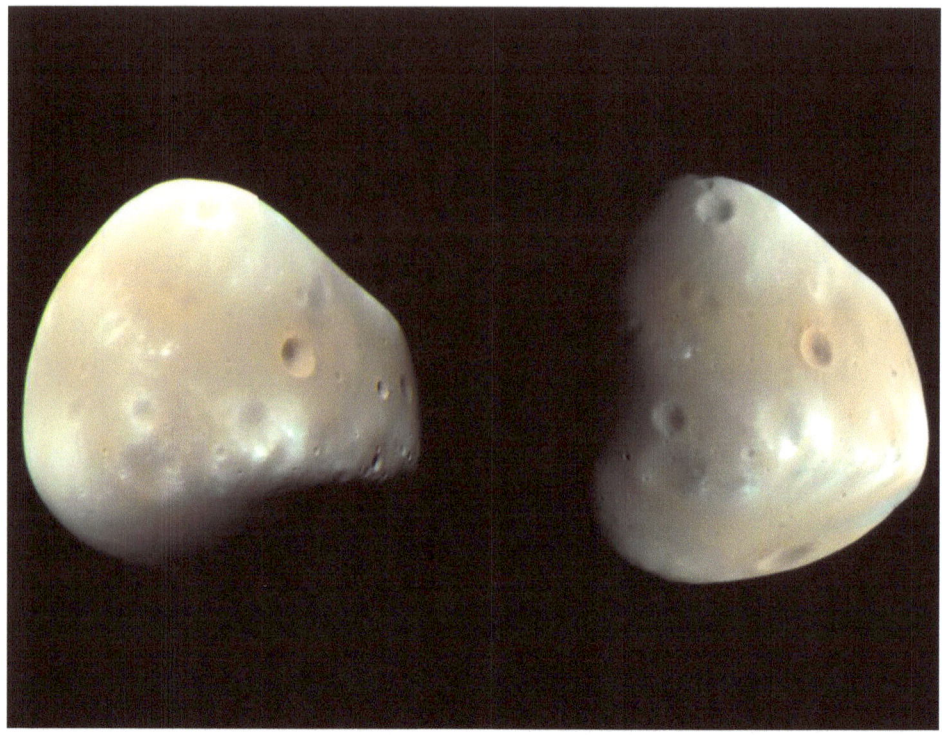

Deimos from the Global Explorer April 2003.

Dione

(Saturn IV; eighteenth moon outward from Saturn)

Albedo: 0.92

Apparent Magnitude: +10.4 at planet's mean opposition

Axis Tilt: 0

Density: 1.478 (water = 1)

Diameter: 1129 by 1122.6 by 1119 kilometers; 700 by 697.4 by 685 miles

Discovery: March 21, 1684 by Giuseppe D. Cassini

Escape Velocity: 0.51 kilometer/second; 1600 feet/second

Mass: 1.095452×10^{21} kilograms

Orbit
    Apoapsis: 379,500 kilometers; 235,810 miles
    Periapsis: 350,000 kilometers; 217,500 miles
    Semi-Major Axis: 377,396 kilometers; 234,363 miles
    Eccentricity: 0.0022
    Inclination to planet's equator: 0.019 degree
    Period: 2.736915 Earth days

Rotation: Synchronous with orbit

Surface Gravity: 0.232 meters/second; 0.024g

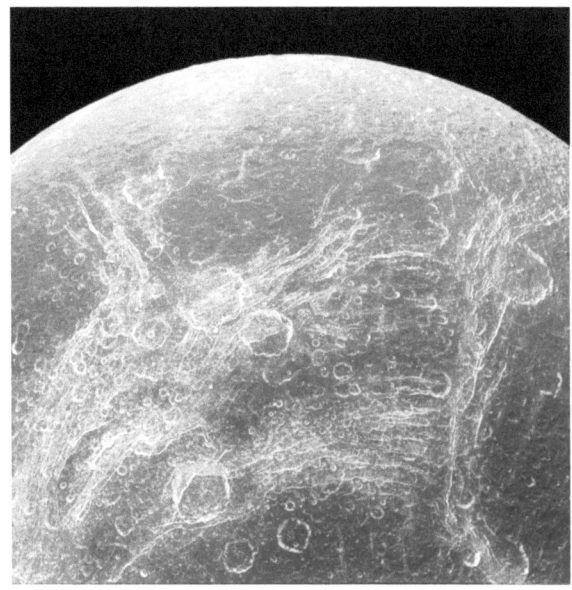

Dione on April 11, 2015 from 68,000 miles/110,000 kilometers showing its unique pattern of chasms.

Earth

(third planet from Sun; one moon)

Albedo: 0.367

Atmosphere

78.08% N2; 20.95% O2; 0.93% Argon; 0.039% CO2. Water is extremely variable, but around 1%.
Surface pressure: 14.7 pounds/square inch; 1 bar

Axis Tilt: 23.439281 degrees

Density: 5.514 This is the largest density of any object in the Solar System, so far as is known. (water=1)

Diameter

Equatorial: 12,756.2 kilometers; 7926.4 miles
Polar: 12,713.6 kilometers; 7900 miles

Escape Velocity: 11.186 kilometers/second, or 6.96 miles per second

This should not be confused with the velocity needed to place into an orbit around Earth, 5 miles/second; 8.05 kilometers/second.

Mass: 5.97237 X 10^24 kilograms; 6.592 X 10^21 tons

Orbit

Aphelion: 151,930,000 kilometers; 1.01559 Astronomical Units; 94,404,988 miles
Semi-major Axis: 149,597,870 kilometers; 1.00000 Astronomical Units; 92,955,807 miles
Perihelion: 147,095,000 kilometers; 0.9832687 Astronomical Units; 91,400,536 miles
The difference between perihelion and aphelion leads to a difference of 6.9% in the energy received from the Sun.
Eccentricity: 0.01671123
Inclination of Orbit to Solar Equator: 7.155 degrees
Mean Orbital Speed: 29.8 kilometers/second; 18.5 miles/second

Rotation

23 hours 56 minutes 4.0989 seconds
speed of rotation at equator: 1674.4 kilometers/hour; 1040.4 miles/hour. Multiply this number by the sine of the latitude for other locations.

Surface Gravity: 9.807 meters/second^2; 1g

Spacecraft: The first spacecraft to return crude photos of Earth was Vanguard 3. Spacecraft photographing Earth may include weather, mapping, spy, and resource satellites.

ECLIPSES

LUNAR

In the 5000 year period from 2000 B.C. to 3000 A.D. there are 12,186 lunar eclipses seen from Earth.
- Penumbral: 4468 (36.7%)
- Partial: 4213 (34.6%)
- Total: 3505 (28.8%)

Penumbral eclipses are usually barely noticeable from Earth. An astronaut on the Moon would see a partial solar eclipse.

SOLAR

In the 5000 year period from 2000 B.C. to 3000 A.D. there are 11,898 solar eclipses seen from Earth.
- Partial: 4200 (35.3%)
- Annular: 3956 (33.2%)
- Total: 3173 (26.7%)
- Hybrid: 569 (4.8%)

Hybrid eclipses are annular for part of their path and total for part of their path.

There must be a minimum of two solar eclipses of some kind in any calendar year.

OTHER PLANETS
- Mars averages 3.4 lunar and 3.4 solar eclipses per Martian day, 3.2 involving Phobos, the remainder involving Deimos.
- Jupiter has from four to seven eclipses per Jovian day involving its four small innermost moons, and typically one to three involving the four large Galilean moons. Its remaining moons rarely are involved in an eclipse, having orbits inclined to Jupiter's orbit by too large an amount, and rarely at a node. The Galilean moons also eclipse and occult one another.
- Saturn has an average of one or two eclipses a week near its nodes. Its inclination prevents more frequent eclipses despite the many moons.
- Uranus has its main moons oriented to its equator, which is inclined so much to its orbit (97.9 degrees) that eclipses are possible only for two brief periods in its 84 year orbit.
- Neptune's axial tilt and the orbital inclinations of its moons limit eclipses to a few per Neptune year.
- Pluto can have eclipses for several months around the time of node crossing (which occur about every 124 Earth years).

Enceladus

(fourteenth moon outward from Saturn; Saturn II)

Albedo: 0.98

Apparent Magnitude: +11.7 at planet's mean opposition

Atmosphere
　91% H_2O; 4% N_2; 3.2% CO_2; 1.7% CH_4
　Pressure: 10^{-6} Earth

Axis Tilt: 1.5 degrees

Density: 1.609 (water = 1)

Diameter: 318.6 by 312 by 308.6 miles; 513.2 by 502.8 by 496.6 kilometers; 0.0395 Earth

Discovery: August 28, 1789 by William Herschel

Escape Velocity: 0.239 kilometer/second; 0.15 mile/second

Mass: 1.08022×10^{22} kilograms

Orbit
　Apoapsis: 239,066 kilometers; 148,461 miles
　Periapsis: 236,830 kilometers; 147,072 miles
　Semi-major axis: 237,948 kilometers; 147,767 miles
　Eccentricity: 0.0047
　Inclination: 0.019 degree to planet's equator
　Period: 1.370218 Earth days; 32.9 hours
　Average Orbital Speed: 10.94 kilometers/second; 6.8 miles/second

Rotation: Synchronous to orbit

Surface Gravity: 0.113 meter/second^2; 0.0114g

Spacecraft
　Voyager 1 ....November 12, 1980 ....fly-by
　Voyager 2 ....August 26, 1981.................fly-by
　Cassini.................February 17, 2005...............orbit planet, fly-by moon

Enceladus on July 27, 2015 from the Cassini probe, 112,000 kilometers; 70,000 miles.

Eris

(dwarf planet; one known moon; largest Solar System object not yet visited by spacecraft)

Albedo: 0.96

Apparent Magnitude: +18.7

Density: 2.52 (water = 1)

Diameter: 2326 kilometers; 1445 miles

Discovery: January 5, 2005 by Michael Brown, Chadwick Trujillo and David Rabinowitz

Escape Velocity: unknown

Mass: 1.66×10^{22} kilograms

Orbit
    Aphelion: 14,602,000,000 kilometers; 9,077,228,000 miles; 97.651 Astronomical Units
    Perihelion: 5,723,000,000 kilometers; 3,524,050,000 miles; 37.911 Astronomical Units
    Semi-Major Axis: 10,166,000,000 kilometers; 6,300,640,000 miles; 67.781 Astronomical Units
    Eccentricity: 0.44068
    Inclination to ecliptic: 44.0445 degrees
    Period: 558.04 years; 202,830 days
    Average Orbital Speed: 3.4338 kilometers/second; 2.134 miles/second

Rotation: 25.9 hours

Surface Gravity: 0.82 meter/second^2; 0.084g

Eros

(first asteroid to be landed upon; first discovered asteroid not limited to Main Belt)

Albedo: 0.25

Apparent Magnitude: +7 to +15

Density: 2.67 (water = 1)

Diameter: 39.75 by 11.3 by 11.3 kilometers; 24.7 by 7 by 7 miles

Discovery: August 13, 1898 by Auguste Charlois and Carl Gustav Witt

Escape Velocity: 10.2 meters/second; 40 feet/second

Orbit
    Aphelion: 1.78245 Astronomical Units; 266,651,000 kilometers; 165,689,000 miles
    Perihelion: 1.3322 Astronomical Units; 199,294.000 kilometers; 123,836,000 miles
    Semi-Major Axis: 1.7824878 Astronomical Units; 269,014,190 kilometers; 163,157,670 miles
    Eccentricity: 0.22267
    Inclination to ecliptic: 10.8287 degrees
    Period: 1.76 years; 643.219 days

Rotation: 5.27 hours

Surface Gravity: 0.0006g

Spacecraft
    NEAR-Shoemaker ....February 14, 2000............orbit/lander

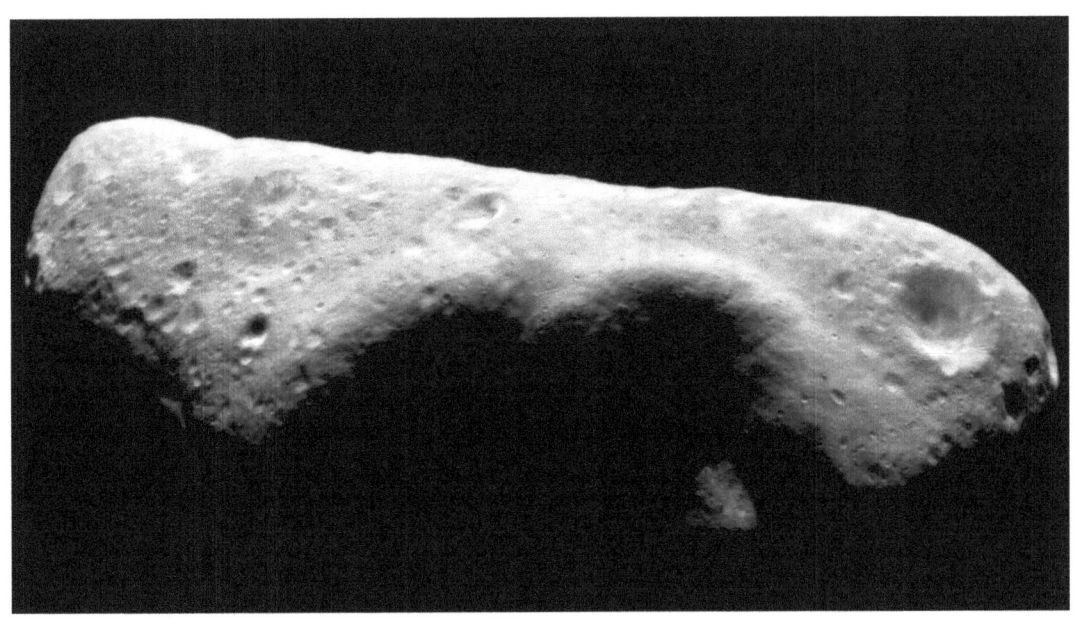

Eros on November 30, 2000, two months before landing.

Europa
(sixth moon outward from Jupiter; Jupiter II)

Albedo: 0.67

Apparent Magnitude: +5.29 at planet's mean opposition

Atmosphere: O2 at a pressure of 10^{-12} bar (Earth's atmosphere at sea level = 1 bar)

Axis Tilt: 0.1 degree

Density: 3.013 (water = 1)

Diameter: 3121.6 kilometers; 1961 miles

Discovered: January 8, 1610 by Galileo Galilei

Escape Velocity: 2.025 kilometers/second; 1.26 miles/second

Mass: 4.8 10^{22} kilograms

Orbit
    Apoapsis: 676,938 kilometers; 420,380 miles
    Periapsis: 412,400 miles; 664,862 kilometers
    Semi-major Axis: 670,900 kilometers; 415,970 miles
    Eccentricity: 0.009
    Inclination: 0.470 degree to planet's equator
    Period: 3.551181 Earth days; 3 days 13 hours 13.7 minutes
    Average Orbital Speed: 13.740 kilometers/second; 8.54 miles/second

Rotation: Synchronous with orbit

Surface Gravity: 1.314 meters/second^2, 0.133g

Spacecraft

Spacecraft	Year	Type
Pioneer 10	1973	fly-by
Pioneer 11	1974	fly-by
Voyager 1	1979	fly-by
Voyager 2	1979	fly-by
Galileo	1995	orbit planet
New Horizons	2007	fly-by

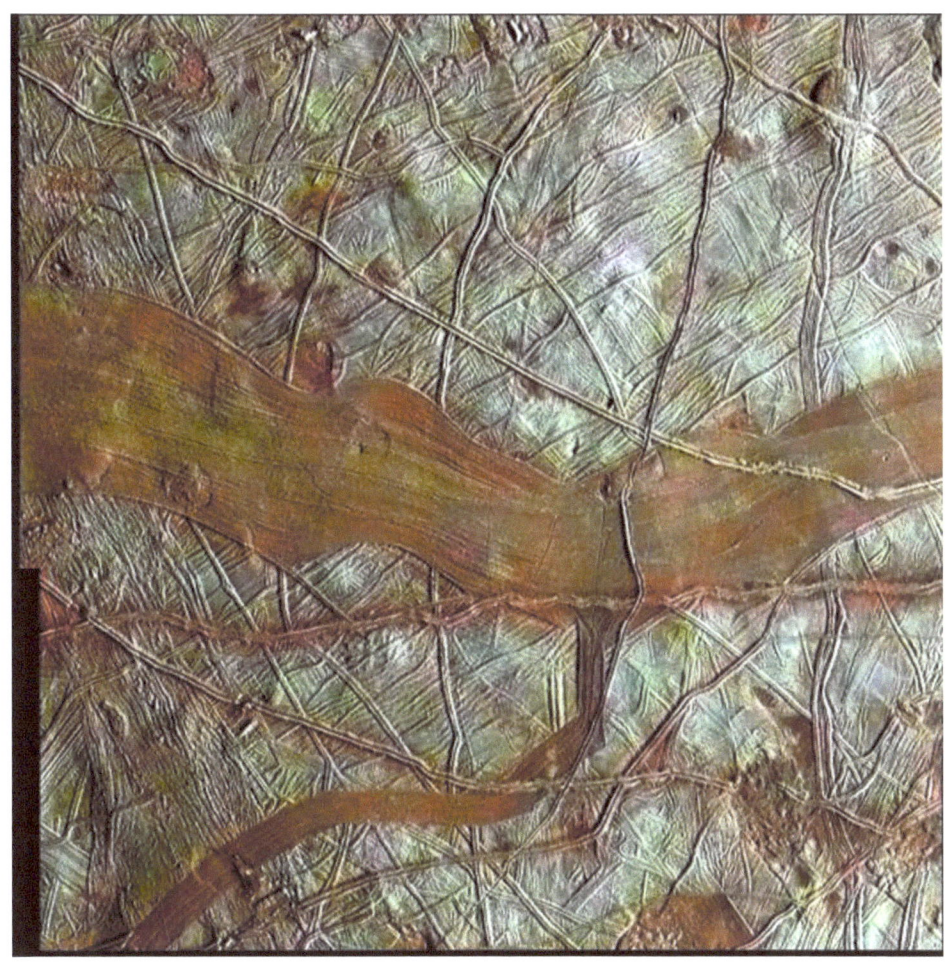

Europa on November 6, 1997 from 13,237 miles/21,700 kilometers.

EXOPLANETS
NEAREST EARTH

The following list shows the *Names* of the nearest stars known to have planets, the *Mass* of such planets in units of Jupiter masses, *Semi-major Axis* of the planets' orbits in Astronomical Units, the *Period* of the planetary orbit in days, the *Year* of discovery, *Right Ascension* and *Declination*.

Name	Mass	SM.Axis	Period	Year	Rt.Ascen	Declin.
Alpha Centauri B	0.0036	0.04	3.24	2012	14h39m36s	-60^50'02"*
Luhman	16	AB	~10	2013	10h49m19s	-53^19'10»
Epsilon Eridani	1.55	3.34	2500	2000	03h32m56s	-09^27'30»
Epsilon Eridani	~0.1	<40	~100,000	2002	03h32m56s	-09^27'30»
Groombridge 34	0.0168	0.0717	11.4	2014	00h18m23s	+44^01'23»
Epsilon Indi	0.97	8.57	>10	2002	22h03m22s	-56^47'10"
Epsilon Indi	0.0063	0.105	14.0	2012	22h03m22s	-56^47'10"
Epsilon Indi	0.0098	0.195	35.4	2012	22h03m22s	-56^47'10"
Tau Ceti	0.011	0.374	94.1	2012	01h44m04s	-15^56'15"
Tau Ceti	0.0135	0.552	168	2012	01h44m04s	-15^56'15"
Tau Ceti	0.021	1.35	642	2012	01h44m04s	-15^56'15"
Kapteyn's Star	0.015	0.168	48.6	2014	05h11m41s	-45^01'06»
Kapteyn's Star	0.022	0.311	122	2014	05h11m41s	-45^01'06»
Gliese 687	0.41	0.058	38.1	2014	17h36m26s	+68^20'21»
LHS 292	0.35	0.04	4.69	2007	10h48m13s	-11^20'14»

* *existence in doubt*

Ganymede

(seventh moon from Jupiter; largest moon in Solar System; Jupiter III)

Albedo: 0.43

Apparent Magnitude: +4.6 at planet's mean opposition

Axial Tilt: 0.33 degrees

Density: 1.936 (water = 1)

Diameter: 5268 kilometers; 3273 miles

Discovered: January 7, 1610 by Galileo Galilei; 365 B.C. by Gan De

Escape Velocity: 2.741 kilometers/second; 1.703 miles/second

Mass: 1.4819×10^{23} kilograms

Orbit
    Apoapsis: 665,420 miles; 1,071,600 kilometers
    Periapsis: 663,980 miles; 1,069,200 kilometers
    Semi-major Axis: 1,070,700 kilometers; 665,300 miles
    Eccentricity: 0.0013
    Inclination: 0.20 degree to planet's equator
    Period: 7 days 3 hours 42.6 minutes; 7.15455296 Earth days
    Average Orbital Speed: 10.88 kilometers/second; 6.76 miles/second

Rotation: Synchronous with orbit

Surface Gravity: 1.428 meters/second^2; 0.15g

Spacecraft
    Pioneer 10 1973
    Pioneer 11 1974
    Voyager 1 1979
    Voyager 2 1979
    Galileo............................2002

Ganymede, the Solar System's largest moon, from orbit around Jupiter.

Gaspra
(first asteroid visited by a spacecraft)

Absolute Magnitude: +11.46

Albedo: 0.22

Axial Tilt: ?

Density: 2.7 (water = 1)

Diameter: 18.2 by 10.5 by 8.9 kilometers; 11.3 by 6.5 by 5.5 miles

Discovery: July 30, 1916 by Grigorii Neujmin

Escape Velocity: 20 feet/second; 6 meters/second

Orbit
   Aphelion: 2.592755 Astronomical Units; 378,894,800 kilometers; 235,434,300 miles
   Perihelion: 1.82668 Astronomical Units; 273,267,400 kilometers; 163,800,500 miles
   Semi-Major Axis: 2.209719 Astronomical Units; 330,569,300 kilometers; 205,406,200 miles
   Eccentricity: 0.173342
   Inclination: 4.10256 degrees
   Period: 1199.79 days; 3.28 years

Spacecraft
   Galileo..............................October 29, 1991 ....fly-by

Gaspra during the 1991 fly-by.

Hyperion
(twenty-third moon outward from Saturn; Saturn VII)

Albedo: 0.25

Apparent Magnitude: +14.1 at planet's mean opposition

Axis Tilt: swings erratically and unpredictably

Density: 0.544 (water = 1)

Diameter: 360.2 by 266 by 205.4 kilometers; 223.8 by 165.3 by 127.6 miles

Discovery: September 16, 1848 by William and George Bond

Escape Velocity: ~50 meters/second; ~160 feet/second (varies with location)

Mass: 5.6199 X 10^9 kilograms

Orbit
    Apoapsis: 1,663,182 kilometers; 1,033,453 miles
    Periapsis: 1,298,936 kilometers; 807,059 miles
    Semi-Major Axis: 1,481,009 kilometers; 920,256 miles
    Eccentricity: 0.1230061
    Inclination: 0.43 degree
    Period: 21.276 Earth days
    Average speed in orbit: 4.44 kilometers/second; 3 miles/second

Rotation: erratic and unpredictable

Surface Gravity: 0.019 meters/second^2; 0.0022g

Spacecraft
    Cassini.............................September 25, 2005

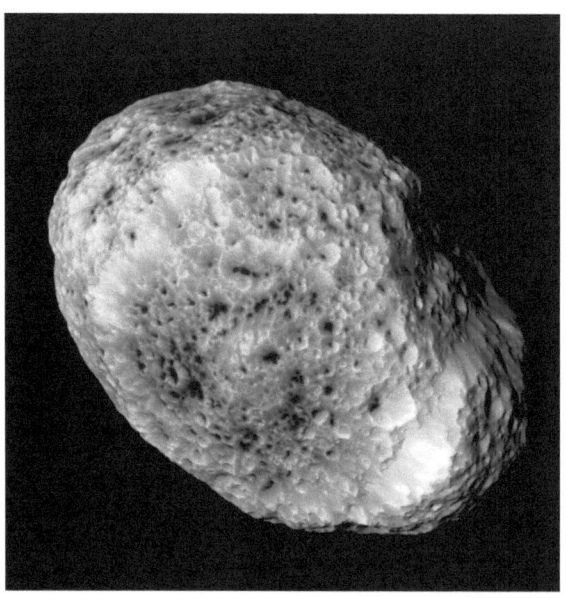

Hyperion from the Cassini orbiter.

Iapetus

(Saturn VIII, twenty-fourth moon outward from Saturn)

Albedo: East hemisphere 0.05; West hemisphere 0.52

Apparent Magnitude: East +11.9; West +10.2 at planet's mean opposition

Axis Tilt: 0

Density: 1.088 (water = 1)

Diameter: 1492 by 1492 by 1424 kilometers; 926.5 by 926.5 by 884.3 miles

Discovery: October 25, 1671 by Giovanni D. Cassini

Escape Velocity: 0.573 kilometer/second; 1879 feet/second

Mass: 1.805635×10^{21} kilograms

Orbit
    Apoapsis: 3,662,704 kilometers; 2,275,900 miles
    Periapsis: 3,458,936 kilometers; 2,149,283 miles
    Semi-major Axis: 3,560,820 kilometers; 2,212,591 miles
    Eccentricity: 0.0286125
    Inclination to planet's equator: 15.47 degrees
    Period: 79.3215 Earth days
    Average orbital speed: 3.26 kilometers/second; 2 miles/second

Rotation: synchronous to orbit

Surface Gravity: 0.223 meter/second^2; 0.024g

Spacecraft
    Pioneer 11
    Cassini

Ida

(asteroid, one moon)

Albedo: 0.2383

Absolute Magnitude: +9.94

Axis Tilt: small

Density: <3.2

Diameter: 31.4 kilometers; 19.5 miles

Discovery: September 29, 1884 by Johann Palisa

Escape Velocity: 20 meters/second; 65 feet/second

Mass: ~4 X 10^16 kilograms

Orbit
    Aphelion: 2.9814 Astronomical Units; 446,011,100 kilometers; 277,138,400 miles
    Perihelion: 2.7450 Astronomical Units; 410,646,000 kilometers; 255,163,700 miles
    Semi-Major Axis: 2.862 Astronomical Units; 428,149,100 kilometers; 266,039,500 miles
    Eccentricity: 0.0452
    Inclination: 1.138 degrees
    Period: 1768.69 days; 4.84 years

Rotation: 4.63 hours

Surface Gravity: 0.0004g

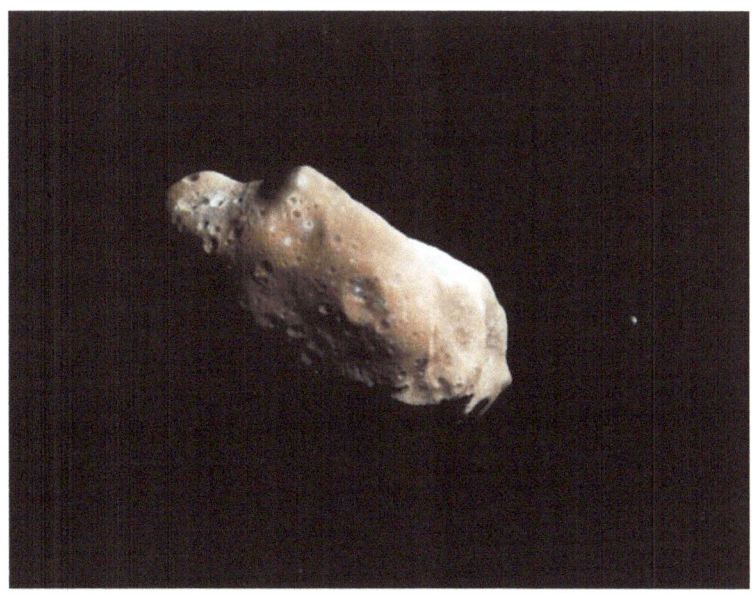

Asteroid Ida with its moon Dactyl.

Io

(fifth moon from Jupiter; densest moon in Solar System; Jupiter I)

Albedo: 0.63

Apparent Magnitude: +5.02 at planet's mean opposition

Atmosphere

Composition: SO_2 90%; SO; NaCl; S; O.

Pressure: 0.02 of Earth

Axis Tilt: small

Density: 3.5275 (water = 1)

Diameter: 3660.0 by 3637.4 by 3630.6 kilometers; 2263.4 miles

Discovered: January 8, 1610 by Galileo Galilei

Escape Velocity: 2.558 kilometers/second; 1.59 miles/second

Mass: 8.932 X 10^{22} kilograms; 0.015 Earth

Orbit

    Apoapsis: 423,400 kilometers; 263,100 miles
    Periapsis: 420,000 kilometers; 261,000 miles
    Semi-major Axis: 421,700 kilometers; 262,000 miles
    Eccentricity: 0.0041
    Inclination: 0.05 degree to planet's equator
    Period: 1.769137786 Earth Days; 42.45930686 hours; 1 day 18 hours 27.6 minutes
    Average Orbital Speed: 17.334 kilometers/second; 10.77 miles/second

Rotation: Synchronous with orbit

Surface Gravity: 1.796 meters/second^2; 0.183g

Spacecraft

    Pioneer 10 ....fly-by
    Pioneer 11 ....fly-by
    Voyager 1 ....fly-by
    Voyager 2 ....fly-by
    Galileo ....orbiter of planet

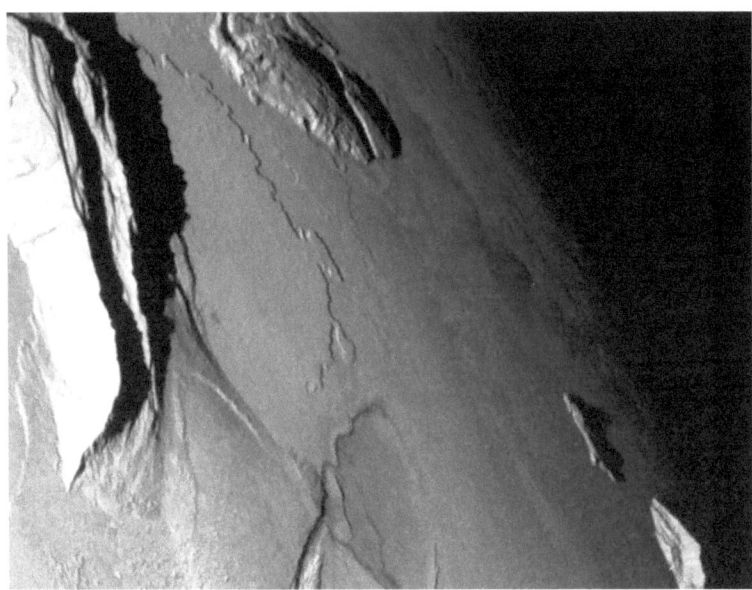

Io February 2000, with the ridge on the left 23,000 feet/6 kilometers high.

Itokawa

(first asteroid with a sample returned to Earth)

Albedo: 0.53

Apparent Magnitude: +17 at mean opposition

Density: 1.92 (water = 1)

Diameter: 1.01 by 0.61 by 0.4 kilometer; 3325 by 2000 by 1300 feet; 0.63 by 0.379 by 0.246 mile

Discovery: September 26, 1998 by Lincoln Near Earth Asteroid Research (LINEAR)

Escape Velocity: 20 centimeters/second; 8 inches/second

Mass: 3.51×10^{10} kilograms

Orbit
    Aphelion: 1.6951 Astronomical Units; 253,583,000 kilometers; 157,569,400 miles
    Perihelion: 0.9531 Astronomical Units; 142,581,700 kilometers; 88,596,200 miles
    Semi-Major Axis: 1.3241 Astronomical Units; 198,082,500 kilometers; 123,082,800 miles
    Eccentricity: 0.26018
    Inclination to ecliptic: 1.6216 degrees
    Period: 1.52 years; 556.52 days

Rotation: 12.13 hours

Surface Gravity: 0.00002g

Spacecraft
    Hayabusa..........................2005........................orbit, sample retrieval

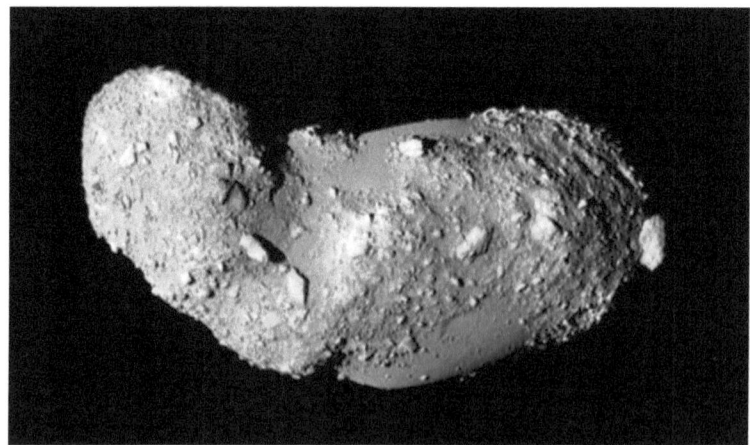

Asteroid Itokawa from the JAXA spacecraft Hiyabusa in 2005.

Janus

(eighth moon outward from Saturn; Saturn X)

Albedo: 0.72

Apparent Magnitude: +14.4

Axial Tilt: 0

Density: 0.63 (water = 1)

Diameter: 203 by 185 by 152.6 kilometers; 126 by 115 by 95 miles

Discovery: December 15, 1966 by Audouin Dollfus

Mass: 1.8975 X 10^{18} kilograms

Orbit
    Apoapsis: 96,064 miles; 154,600 kilometers
    Periapsis: 94,033 miles; 151,133 kilometers
    Semi-Major Axis: 151,460 kilometers; 95,197 miles
    Eccentricity: 0.0068
    Inclination: 0.163 degree
    Period: 0.644160 day

Rotation: Synchronous

Spacecraft
    Pioneer 11 ....September 1, 1979
    Cassini............................2012

Surface Gravity: 0.011 meter/second^2; 0.0013g

Jupiter
(fifth planet from Sun, largest planet, 67 moons [most] + rings)

Albedo: 0.52

Angular Diameter: 29.8 to 50.1 arc seconds

Apparent Magnitude: -1.6 to -2.94

Axis Tilt: 3.13 degrees

Composition: 89.8% H2; 10.2% He; 0.3% CH4; 0.026% NH3; 0.003% HD; 0.0006% C2H6. Clouds of NH4SH; NH3; PH3.

Density: 1.326 (water = 1)

Diameter
  Equatorial: 148,784 kilometers; 92,450 miles; 11.209 Earth
  Polar: 133,708 kilometers; 83,082 miles; 10.517 Earth

Escape Velocity: 59.5 kilometers/second; 37 miles/second

Mass: 1.8986 X 10^27 kilograms; 317.8 X Earth; 1/1047 Sun.

Orbit
  Aphelion: 816,520,800 kilometers; 507,362,680 miles; 5.458104 Astronomical Units
  Perihelion: 740,573,600 kilometers; 460,171,100 miles; 4.950429 Astronomical Units
  Semi-major Axis: 778,547,200 kilometers; 468,766,840 miles; 5.204267 Astronomical Units
  Eccentricity: 0.048775
  Inclination: 1.305 degrees
  Average orbital speed: 13.07 kilometers/second; 8.2 miles/second
  Orbital Period: 11.8618 Earth years; 4332.59 days; 10,475.89 Jovian days
  Synodic Period: 398.88 days

Rings
  Discovery: 1979 by Voyager 1

Halo
  Distance: 92,000 to 122,500 kilometers; 57,200 to 76,100 miles

Main
  Distance: 122,500 to 129,000 kilometers; 76,100 miles to 80,200 miles.

Amalthea Gossamer
  Distance: 129,000 to 182,000 kilometers; 80,200 to 113,100 miles

Thebe Gossamer
  Distance: 150,000 to 226,000 kilometers; 93,200 to 140,000 miles

Rotation: 9.925 hours = 9 hours 59 minutes 30 seconds

    Rotational Speed at Equator: 12.6 kilometers/second; 7.83 miles/second

Surface Gravity: 24.79 meters/second^2; 2.5g (Note: Jupiter has no solid surface, these figures are for an arbitrary point in the upper atmosphere.)

Spacecraft

Spacecraft	*Date*	*Result*
Pioneer 10	December 12, 1973	fly-by
Pioneer 11	December 4, 1974	fly-by
Voyager 1	March 5, 1979	fly-by
Voyager 2	July 9, 1979	fly-by
Ulysses	February 8, 1992	fly-by
Galileo	December 7, 1995	orbiter/impact
Cassini	December 30, 2000	fly-by
New Horizons	February 28, 2007	fly-by

All are from the United States.

LIGHTYEAR

A lightyear is the distance light travels in a vacuum in one year.
    63,241.1 Astronomical Units
    9,460,730,472,580.8 kilometers
    5,878,625,373,183.608 miles
    0.306391 parsec

LOCAL GROUP

This term refers to galaxies in the same cluster of galaxies as the galaxy we live in, known as the Milky Way. The listing is arranged by *name*/designation, *distance* in lightyears, *type* of galaxy, and *apparent magnitude*.

NAME	DISTANCE	TYPE	APP. MAG.
Milky Way	0	SB	-
Sagittarius Dwarf	75,000	dE7	+4.5
Ursa Major II Dwarf	98,000	dSpherical	+14.3
Large Magellanic Cloud	163,000	Irr	+0.9
Bootes I	197,000	dSpherical	+13.1
Small Magellanic Cloud	206,000	SBpec	+2.7
Ursa Minor I Dwarf	206,000	E4	+11.9
Draco Dwarf	258,000	E0pec	+10.9
Sextans Dwarf	281,000	dSpherical	+12
Ursa Major I Dwarf	330,000	dSpherical	+13.4
Carina Dwarf	330,000	dE3	+11.3
Leo II Dwarf	701,000	E0pec	+12.45
Leo I Dwarf	820,000	E3	+11.5
Fornax Dwarf	960,000	dSpherical	+9.3
Phoenix Dwarf	1,440,000	Irr	+13
Barnard's Galaxy	1,630,000	Irr	+9.32

Name	Distance (ly)	Type	Magnitude
NGC 185	2,010,000	dE3pec	+10
Andromeda II	2,130,000	dE0	+15.1
IC 10	2,200,000	dIrr	+12.2
NGC 147	2,200,000	dE5pec	+10.36
Leo A	2,250,000	I/B	+12.9
IC 1613	2,350,000	I/B	+9.9
Andromeda I	2,430,000	dE3pec	+13.9
Andromeda III	2,440,000	dE2	+15.2
Cetus Dwarf	2,460,000	dSpec	+14.4
M32	2,480,000	E2	+8.73
Andromeda IX	2,500,000	dE	+16.3
LGS 3	2,510,000	dIrr	+16.2
Andromeda V	2,520,000	dSpec	+16.7
Pegasus Dwarf	2,550,000	dSpec	+14.1
M31 (Andromeda)	2,560,000	SA	+4.17
M33 (Triangulum)	2,640,000	SA	+6.19 (most distant object visible to eyes)
M110	2,690,000	E6	+8.7
Andromeda VIII	2,700,000	dSpec	+9.1

Globular clusters are not included in this list.

Lutetia
(asteroid)

Albedo: 0.19

Apparent Magnitude: +10.8 at mean opposition

Axial Tilt: 96 degrees

Density: 3.4 (water = 1)

Diameter: 59 by 66 miles;

Discovery: November 15, 1852 by Hermann Goldschmidt

Escape Velocity: 0.069 kilometer/second; 0.043 mile/second

Mass: 1.7×10^{18} kilograms

Orbit
    Aphelion: 2.835795 Astronomical Units; 424,228,900 kilometers; 263,603,600 miles
    Perihelion: 2.0339907 Astronomical Units; 304,280,700 kilometers; 189,071,200 miles
    Semi-Major Axis: 2.434893 Astronomical Units; 364,254,300 kilometers; 226,337,400 miles
    Eccentricity: 0.164
    Inclination: 3.064 degrees
    Period: 1387.9 days; 3.80 years
    Average speed: 18.96 kilometers/second; 11.78 miles/second

Rotation: 8.1655 hours; 0.3402 day

Spacecraft
    Rosetta............................July 10, 2010..........fly-by

Surface Gravity: 0.049g

MAGNITUDE SCALE

A difference in magnitude	Is a difference in brightness of
1	2.51188643
2	6.30957344
3	15.8489319
4	39.8107171
5	100.000000
6	251.188643
7	630.957344
8	1584.89319
9	3981.07171
10	10,000.0000
11	25,118.8643
12	63,095.7344
13	158,489.319
14	398,107.171
15	1,000,000.000

Mars

(fourth planet from Sun; two moons)

Albedo: 0.17

Apparent magnitude: +1.6 to -3.0

Atmosphere: 95.97% CO_2; 1.93% Argon; 1.89% N_2; 0.146% O_2; 0.055% CO

Pressure: 0.008% of Earth

Axis Tilt: 25.19 degrees

Density: 3.9335 (water = 1)

Diameter: 6779 kilometers; 4260 miles; 0.533 Earth

Escape Velocity: 5.027 kilometers/second, 3.03 miles/second

Mass: 6.4171 X 10^{23} kilograms; 0.107 Earth

Orbit
    Aphelion: 249.2 million kilometers; 154,864,000 miles; 1.6660 Astronomical Units
    Perihelion: 206.7 million kilometers; 128,410,000 miles; 1.3814 Astronomical Units
    Semi-Major Axis: 227,939,100 kilometers; 141,635,000 miles; 1.523679 Astronomical Units
    Eccentricity: 0.0935
    Inclination: 1.85 degrees
    Period: 1.8808 Earth years; 686.971 Earth days; 668.5991 Mars days
    Synodic period: 779.96 days
    Average orbital speed: 24.077 kilometers/second; 14.96 miles/second
    Nearest Earth: August 27, 2003-- 55,758,006 kilometers; 34,646,419 miles; 0.37271925 Astronomical Units

Rotation: 24 hours 37 minutes 22 seconds; 1.025957 Earth days

Solar day: 24 hours 39 minutes 35.244 seconds
    Equatorial rotation speed: 868.22 kilometers/hour; 241.27 meters/second; 539.5 miles/hour

Surface Gravity: 3.711 meters/second^2; 0.376g

Spacecraft (successful only)

Name	Arrived	Result
Mariner 4	July 14, 1965	fly-by
Mariner 9	November 14, 1971	orbiter
Viking 1	July 20, 1976	orbiter/lander
Viking 2	September 4, 1976	orbiter/lander
Mars Pathfinder	July 4, 1997	rover

Global Surveyor	September 12, 1997	orbiter
Mars Odyssey	October 24, 2001	lander
Mars Express	December 25, 2003	orbiter
Spirit	January 4, 2004	rover
Opportunity	January 25, 2004	rover
Reconnaissance	March 10, 2006	orbiter
Rosetta	February 25, 2007	fly-by
Dawn	February 17, 2009	fly-by
Curiosity	August 6, 2012	rover
MAVEN	September 22, 2014	orbiter
Mangalyaan	September 24, 2014	orbiter

Mars Express and Rosetta are from the European Space Agency. Mangalyaan is from the Indian Space Research Organization. All others are from the United States.

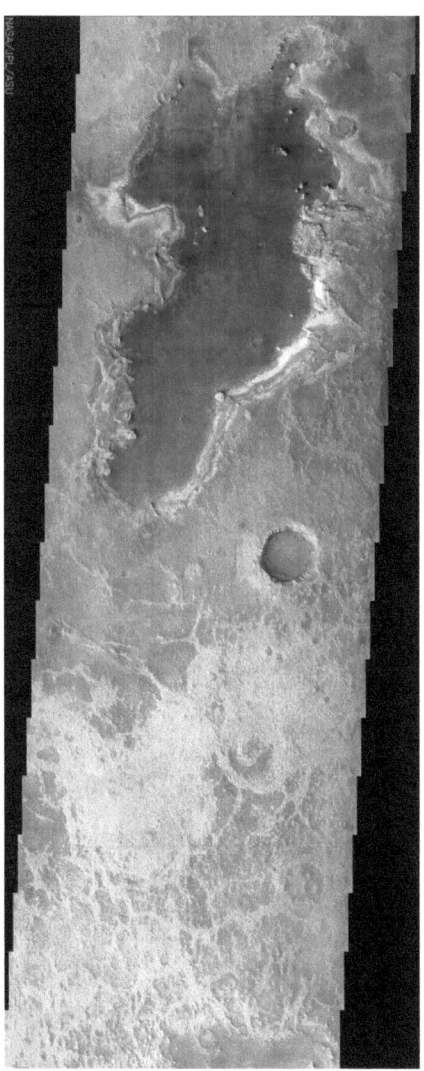

A portion of Mars' surface from Mars Odyssey showing craters and other features.

Mathilde
(asteroid)

Albedo: 0.0436

Absolute Magnitude: +10.3

Axial Tilt: small

Density: 1.3

Diameter: 52.8 kilometers; 32.8 miles

Discovery: November 12, 1885 by Johann Palisa

Orbit
    Aphelion: 3.344619 Astronomical Units; 500,347,900 kilometers; 310,901,800 miles
    Perihelion: 1.942988 Astronomical Units; 290,666,900 kilometers; 180,612,000 miles
    Semi-Major Axis: 2.64304 Astronomical Units; 395,393,200 kilometers; 245,685,900 miles
    Eccentricity: 0.26577
    Inclination: 6.741 degrees
    Period: 1572.38 days; 4.30 years

Rotation: 417.7 hours; 17.4 days

Spacecraft
    NEAR-Shoemaker .... June 27, 1997 .... fly-by

Mercury

(The planet nearest the Sun, and the smallest planet; no moons)

Albedo: 0.142

Apparent Magnitude: -2.6 to +5.1

Axis Tilt: 0.027 degree; 2.04' (least of any planet)

Density: 5.427 (water = 1) This is second highest among the planets, only Earth being (slightly) greater, and that mainly due to compression.

Diameter: 3010 miles; 4879.4 kilometers; 0.3829 Earth

Escape Velocity from Surface: 4.25 kilometers/second; 2.64 miles/second

Mass: 3.3022 X 10^{23} kilograms; 0.055 Earth

Orbit
    Aphelion: 0.466697 Astronomical Units; 69,816,900 kilometers; 43,381,300 miles
    Perihelion: 0.307499 Astronomical Units; 46,001,200 kilometers; 28,583,800 miles
    Semi-Major axis: 0.387098 Astronomical Units; 57,909,050 kilometers; 35,983,100 miles
    Eccentricity: 0.205630
    Inclination to ecliptic: 7.005 degrees
    Period: 87.87 days
    Average Orbital Speed: 47.362 kilometers per second; 29.43 miles per second
    Synodic Period: 115.88 days

Rotation: 3.026 meters/second; 1.88 feet/second

Surface gravity: 3.7 meters/second^2; 0.38g (where 1g = surface gravity of Earth)

Visiting spacecraft
    Mariner 10 (NASA) March 29, 1974 to March 16, 1975 (three flybys)
    MESSENGER (NASA) In orbit January 14, 2008 to April 30, 2015, when it crashed on Mercury's surface

Mercury from 116,000 kilometers/72,000 miles showing crater Abedin.

Metis
(closest moon of Jupiter)

Albedo: 0.061

Apparent Magnitude: +17.5 at Jupiter's mean opposition

Axial Tilt: 0

Density: 0.86 (water = 1)

Diameter: 37 by 24 by 20.5 miles; 60 by 40 by 34 kilometers

Discovered: March 4, 1979 by Stephen Sinnott

Escape Velocity: 0.012 kilometers/second; 480 feet/second

Mass: ~3.6×10^{16} kilograms

Orbit
    apoapsis: 128,026 kilometers; 78,002 miles
    periapsis: 127,974 kilometers; 77,973 miles
    Semi-major axis: 128,000 kilometers; 77,988 miles
    Eccentricity: 0.0002
    Inclination to planet's equator: 0.06 degree
    Period: 0.294780 Earth day; 7 hours 4.5 minutes (shortest orbital period of any known moon, one of just three known to orbit in less than its planet's day)
    Average orbital speed: 31.501 kilometers/second; 19.56 miles/second

Rotation: Synchronous with orbit

Surface Gravity: 0.005 meters/second^2; 0.00051g

Spacecraft
    Voyager 1 ....March 4, 1979
    Galileo............................2002

Mimas

(tenth moon outward from Saturn; Saturn I)

Albedo: 0.96 (highest in Solar System)

Apparent Magnitude: +12.8 at planet's mean opposition

Axial Tilt: 0

Density: 1.15 (water = 1)

Diameter: 396 kilometers; 246 miles; 0.0311 Earth

Discovery: September 17, 1789 by William Herschel

Escape Velocity: 0.159 kilometers/second; 0.099 mile/second

Mass: 3.7493 X 10^{14} kilograms

Orbit
    Apoapsis: 183,176 kilometers; 117,478 miles
    Periapsis: 181,902 kilometers; 112,961 miles
    Semi-Major Axis: 185,539 kilometers; 115,225 miles
    Eccentricity: 0.0196
    Inclination to Planet's Equator: 1.566 degrees
    Period: 0.942422 Earth days
    Average Orbital Speed: 14.28 kilometers/second;
        8.87 miles/second

Rotation: Synchronous to orbit

Surface Gravity: 0.064 meters/second^2; 0.0065g

Spacecraft
    Pioneer 11 ....September 1, 1979 ....fly-by
    Voyager 1 ....1980...................fly-by
    Voyager 2 ....1981...................fly-by
    Cassini..............2004..................orbit planet

Mimas on February 13, 2010.

Miranda

(fourteenth moon outward from Uranus; Uranus V)

Albedo: 0.32

Apparent Magnitude: +15.8 at planet's mean opposition

Axial Tilt: 0

Density: 1.2 (water = 1)

Diameter: 480 by 468.4 by 465.8 kilometers; 300 by 291 by 290 miles

Discovery: February 16, 1948 by Gerard P. Kuiper

Escape Velocity: 0.193 kilometer/second; 0.12 mile/second

Mass: 6.59 X 10^{19} kilograms

Orbit
    Apoapsis: 168,160 kilometers; 104,490 miles
    Periapsis: 90,287 kilometers; 56,102 miles
    Semi-Major Axis: 129,222 kilometers; 80,295 miles
    Eccentricity: 0.0013
    Inclination to planet's equator: 4.232 degrees
    Period: 1.413479 Earth days
    Average Orbital Speed: 6.66 kilometers/second; 4.14 miles/second

Rotation: synchronous with orbit

Surface Gravity: 0.008g

Spacecraft
    Voyager 2 ....January 26, 1986....fly-by

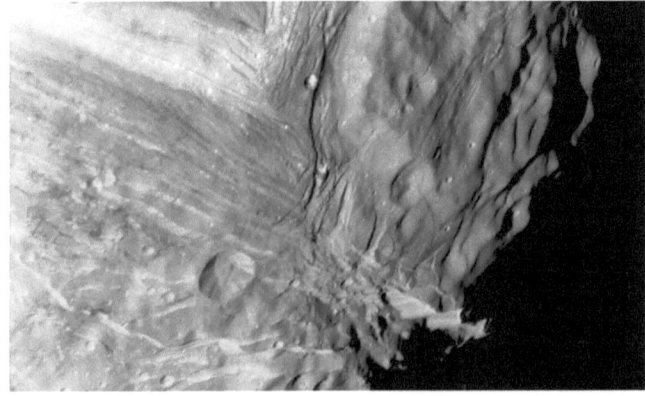

Miranda on January 24, 1986 from 38,000 kilometers/22,000 miles. The crater seen is 25 kilometers/15 miles in diameter.

The Moon

Albedo: 0.072

Apparent Magnitude: maximum at Full phase is -12.9

Density: 3.3464 (Earth = 5.52; water = 1)

Diameter: 2159.9 miles; 3474.2 kilometers; 0.273 Earth

Distance

    Lunar distances are subject to effects from the gravity of the Sun as well as several planets, thus a range is given here for such values as apogee and perigee. It should be noted that the Moon's orbit is always concave towards the Sun, not the Earth.

    Semi-major axis: 384,401 kilometers; 238,464 miles; 0.00256956 Astronomical Units

    Perigee: 356,400 to 370,900 kilometers; 221,500 miles

    Apogee: 404,800 to 406,700 kilometers; 252,700 miles

Escape Velocity: 2.38 kilometers/second; 1.49 miles/second

Period

The Moon's orbital period, regardless of which figure is used (see below) is only an average, as the Sun's gravity, and to a lesser extent the gravities of Venus, Mars, Jupiter, and even other planets, causes variations.

- Anomalistic: 27 days 13 hours 18 minutes 33.2 seconds; 27.554551 days.
  This refers to the lunar period from perigee to perigee. The line connecting perigee and apogee is called the line of apsides, and advances slowly, completing a circuit in about nine years.
- Draconic (also called Nodal): 27 days 5 hours 5 minutes 35.8 seconds; 27.21220 days.
  Based on the Moon returning to the same point in space where its orbit crosses an intersection with the plane of the Earth's orbit (known as the ecliptic). This intersection regresses along the lunar orbit, a complete regression taking 18.6 years.
- Sidereal: 27 days 7 hours 43 minutes 11.5 seconds; 27.321661 days.
  This is the lunar period as seen by an observer at an infinite distance.
- Synodic: 29 days 12 hours 44 minutes 2.9 seconds; 29.530589 days. The word synodic is derived from an Ancient Greek word for a meeting or gathering of priests. It is the only one of the five types of lunar period ever used as the basis of a calendar (although various ancient peoples were acquainted with the draconic month, which determines the timing of eclipses). It is measured from New Moon to New Moon, thus is the only one that keeps pace with lunar phases.
- Tropical: 27 days 7 hours 43 minutes 4.7 seconds; 27.321582 days: The period for the Moon to return to the same position with respect to the Vernal Equinox.
- The draconic period relates to the frequency of eclipses, while the anomalistic period will help determine totality versus annular for solar eclipses. See the *Eclipse* entry.

While all of the above periods can be calculated for the various moons of the other planets, in fact the quoted periods for all of them is only the sidereal period.

Rotation: Synchronous with sidereal period

Speed, Rotational at the Equator: 4.267 meters/second; 14.651 feet/second

Average Orbital Speed: 0.6392 kilometer/second; 0.3972 mile/second; 2097 feet/second

Surface area: 3.3793 X 10^7 square kilometers; 1.30475 X 10^7 square miles; 0.074 Earth

Surface Gravity: 1.622 meters/second^2; 5.32 feet/second^2; 0.16g

Craters: estimated at 600,000 per million square miles of surface area for craters with diameters greater than one kilometer (0.62 mile)

Earth's Moon from the Lunar Reconnaissance Orbiter showing a typical large crater with wall slumping and a central peak.

MOONS

CLOSEST

- Mars: Phobos, 9517.58 kilometers to 9234.42 kilometers; 5914 to 5738 miles. Given the diameter of Mars, Phobos is only 5989 kilometers or 3721 miles above the surface, thus the closest moon in the Solar System.
- Jupiter: Metis, 77,973 miles to 78,003 miles; 48,450 to 48,469 kilometers
- Saturn: S/2009 S1 is all the name this tiny moon has, 72,600 miles; 116,838 kilometers, located in the planet's B ring.
- Uranus: Cordelia at 40,895 miles; 65,814 kilometers.
- Neptune: Naiad at 29,944 to 29,956 miles; 48,190 to 48,210 kilometers.
- Pluto: Charon at 12,160 miles; 19,570 kilometers.
- Haumea: Namaka at an approximate distance of 15,900 miles; 25,600 kilometers.

LARGEST

- Mars: the larger of Mars' moons is Phobos, 27 by 22 by 18 kilometers; 16.8 by 13.67 by 11.2 miles.
- Jupiter: the largest moon is Ganymede, 3273 miles; 5268 kilometers diameter. This is also the largest moon in the entire Solar System. Second largest of Jupiter's moons is Callisto at 2995 miles; 4821 kilometers.
- Saturn: the largest moon is Titan, at 3200 miles; 5152 kilometers diameter. This is also the second largest moon in the Solar System. Second largest of Saturn's moons is Rhea at 948 miles, 1525 kilometers.
- Uranus: the largest moon is Titania, with a diameter of 979 miles; 1575 kilometers. Second largest is Oberon at 946 miles; 1522 kilometers.
- Neptune: the largest moon is Triton at 1680 miles; 2707 kilometers. Second largest is Proteus at 271 by 258 by 250 miles; 436 by 415 by 402 kilometers.
- Pluto: the largest moon is Charon at 1208 kilometers; 750 miles.
- Haumea: the larger moon is Hi'iaka at about 217 miles; 349 kilometers.

MOST DISTANT MOON

- Mars: The more distant of the moons of Mars is Deimos, 14,580 miles; 23,470.9 kilometers. The orbital period is 30 hours 18 minutes; 1.26244 days.
- Jupiter: S/2003 J-2 (no other name yet assigned), 15,270,500 to 22,350,800 miles; 24,575,500 to 35,970,100 kilometers. The orbital period is 981.55 days.
- Saturn: Fornjot, 12,387,000 to 18,097,800 miles; 19,935,000 to 29,125,600 kilometers. The orbital period is 1432.16 days.

- Uranus: Ferdinand, 8,200,000 to 17,758,200 miles; 13,197,000 to 28,622,500 kilometers. The orbital period is 2805 days; 7 years 249 days.
- Neptune: Neso, 13,188,000 to 47,094,000 miles; 21,224,000 to 75,790,000 kilometers; 0.14187 to 0.5066 Astronomical Units. This is the most distant moon known in the Solar System. Not even Jupiter could hold onto a moon this far away. The orbital period is 26 years 250 days; 9740.7 days. Neso makes just 6.5 orbits of Neptune while the planet orbits the Sun once.
- Pluto: Hydra, 40,100 to 40,300 miles; 64,500 to 64,860 kilometers. The orbital period is 38.206 days.

NUMBER OF MOONS

The number of known moons for each planet and dwarf planet:

Mercury	0
Venus	0
Earth	1
Mars	2
Ceres	0
Jupiter	67
Saturn	62
Uranus	27
Neptune	14
Pluto	5
Haumea	2
Orcus	1
Makemake	1
Quaoar	1
Eris	1

Over 200 asteroids are known to have one or more moons.

MOST MOONS DISCOVERED

- Asaph & Angelina Stickney Hall discovered both moons of Mars
- Galileo Galilei, Brett Gladman, and Seth Nicholson each discovered four moons of Jupiter, but the record holder is Scott Shepherd with 42 of Jupiter's moons.
- Brian Marsden was co-discoverer of 25 of Saturn's moons, while Scott Shepherd was co-discoverer of 23, some with Marsden. Giovanni Cassini trails badly with three moons.
- Brett Gladman found nine of the moons of Uranus, followed by Matthew Holman with six.
- William Fraser and T. Grav co-discovered four moons of Neptune, and Matthew Holman independently discovered four other moons of Neptune.
- Mark Showalter discovered two of Pluto's moons, as did the consortium of Hal Weaver, S. Allen Stern, Max Mutchler, Andrew Steffl, Marc Bluie, William Merline, John Spencer, Eliot Young, and Leslie Young.

- Michael Brown, Chadwick Trujillo, and David Rabinowitz discovered both moons of Haumea.
- Michael Brown and T.-A. Suer discovered the moon of Orcus.
- Michael Brown discovered the moon of Quaoar.
- The moon of Eris was discovered by Antonine Bouchez, Michael Brown, Chadwick Trujillo, and David Rabinowitz.

MOUNTAINS, HIGHEST

Object	Mountain	Height
Mercury	Caloris Montes	3 kilometers; 1.9 miles
Venus	Skadi Mons	6.4 kilometers; 4.0 miles
Earth	Mount Everest	8.848 kilometers; 29,029 feet
Moon	Mount Huyghens	5.5 kilometers; 3.4 miles
Mars	Olympus Mons	21.9 kilometers; 14 miles
Vesta	Rheasylvia Central Peak	22 kilometers; 14 miles
Ceres	Ahuna Mons	6 kilometers; 4 miles
Io	Boosaule Montes	18.2 kilometers; 11.3 miles
Mimas	Herschel Central Peak	7 kilometers; 4 miles
Tethys	Scheria Montes	4 kilometers, 2.5 miles
Dione	Janiculum Dorsa	1.5 kilometers; 0.9 mile
Titan	Mithrim Montes	2.0 kilometers; 1.25 miles
Iapetus	equatorial ridge	20 kilometers; 12.5 miles
Oberon	unnamed	~11 kilometers; ~7 miles#
Pluto	Norgay Montes	3.5 kilometers; 2.2 miles

#believed to be the central peak of a large crater located in a portion of Oberon not seen by the spacecraft.

Naiad

(Neptune III, closest moon to Neptune)

Albedo: 0.07

Apparent Magnitude: +23.9 at planet's mean opposition

Axis Tilt: 0

Density: 1.3 (water = 1)

Diameter: 96 by 60 by 52 kilometers; 59 by 37 by 32 miles

Discovery: September 18, 1989

Escape Velocity: 0.028 kilometer/second; 0.017 mile/second

Mass: 1.9×10^{17} kilograms

Orbit
    Apoapsis: 42,244 kilometers; 26,250 miles
    Periapsis: 42,210 kilometers; 26,228 miles
    Semi-Major Axis: 42,227 kilometers; 26,240 miles
    Eccentricity: 0.0004
    Inclination to planet's equator: 4.75 degrees
    Period: 0.2943958 days

Rotation: Synchronous with orbit

Surface Gravity: 0.012 meter/second^2; 0.0009g

Spacecraft
    Voyager 2 .... September 1989 .... fly-by

Neptune

(eighth planet from Sun; 14 moons + rings)

Albedo: 0.41

Apparent Magnitude: +8.0 at mean opposition

Atmosphere: 80% H_2; 19% He; 1% CH_4

Axis Tilt: 28.32 degrees

Density: 1.638 (water = 1)

Diameter

Equatorial: 49,682 kilometers; 30,870 miles; 3.883 Earth

Polar: 48,682 kilometers; 30,250 miles; 3.829 Earth

Discovery: September 23, 1846 by Johann Galle with input by John Couch Adams & Urban Le Verrier

Escape Velocity: 23.5 kilometers/second; 14.6 miles/second

Mass: 1.0243 X 10^{26} kilograms; 17.147 Earth

Orbit
    Aphelion: 4,537,580,900 kilometers; 2,819,522,000 miles; 30.331855 Astronomical Units
    Perihelion: 4,459,504,400 kilometers; 2,771,008,000 miles; 29.809946 Astronomical Units
    Semi-Major Axis: 4,498,542,600 kilometers; 2,795,265,000 miles; 30.070900 Astronomical Units
    Eccentricity: 0.00867797
    Inclination: 1.767975 degrees
    Period: 164.8 years; 60,190.03 Earth days; 89,666.4 Neptune days
    Synodic Period: 367.49 days
    Average Orbital Speed: 5.43 kilometers/second; 3.37 miles/second

Rings

Galle	40,900 to 42,900 kilometers; 25,410 to 26,660 miles
Le Verrier	53,150 to 53,255 kilometers; 32,990 to 35,590 miles
Lassell	53,200 to 57,200 kilometers; 33,025 to 35,540 miles
Arago	57,200 kilometers; 35,540 miles
Adams	62,910 to 62,960 kilometers; 39,090 to 39,120 miles

    Note: Neptune's rings are named for people important in studying or discovering Neptune

Rotation: 0.6713 days; 16 hours 06 minutes 36 seconds

Surface Gravity: 11.15 meters/second^2; 1.14g

Spacecraft
    Voyager 2 .... August 25, 1989 .... fly-by

Nereid

(Neptune II; eighth moon outward from Neptune)

Albedo: 0.155

Apparent Magnitude: +14 at planet's mean opposition

Density: ~1.5

Diameter: 340 kilometers; 211 miles

Discovery: May 1, 1949 by Gerard P. Kuiper

Orbit
    Apoapsis: 9,655,000 kilometers; 5,999,000 miles
    Periapsis: 1,372,000 kilometers; 853,000 miles
    Semi-Major Axis: 5,513,787 kilometers; 3,425,000 miles
    Eccentricity: 0.7507; Largest of any moon in the Solar System
    Inclination: 32.55 degrees
    Period: 360.1362 days
    Average Orbital Speed: 9.34 meters/second; 30.6 feet/second

Rotation: ~11.5 hours?

Spacecraft
    Voyager 2 .... April 20, 1989 .... fly-by planet

Neso

(Neptune XIII; fourteenth moon outward from planet)

Albedo: 0.04

Apparent Magnitude: +19 at planet's mean opposition

Axis Tilt: unknown

Density: 1.5? (water = 1)

Diameter: 60 kilometers; 37 miles

Discovery: August 14, 2002 by Matthew Holman and Brett Gladman

Escape Velocity: unknown

Mass: ~2 X 10^{17} kilograms

Orbit
    Apoapsis: 75,790,000 kilometers; 47,094,000 miles; 0.5066 Astronomical Units; Most distant known moon in Solar System
    Periapsis: 21,224,000 kilometers; 13,188,000 miles; 0.14187 Astronomical Units
    Semi-Major Axis: 0.3295167 Astronomical Units; 49,285,000 kilometers; 30,624,000 miles
    Eccentricity: 0.5714
    Inclination to planet's equator: 136.439 degrees (i.e. retrograde)
    Period: 9740.73 days; 26.67 years (6.5 orbits/Neptune year)

Rotation: unknown

Oberon

(Uranus IV; eighteenth moon outward from Uranus)

Albedo: 0.31

Apparent Magnitude: +14.1 at planet's mean opposition

Axis Tilt: 0

Density: 1.65 (water = 1)

Diameter: 1523 kilometers; 946 miles

Escape Velocity: 0.7272 kilometer/second; 0.452 mile/second

Mass: 3.014×10^{21} kilograms

Orbit
    Apoapsis: 584,337 kilometers; 363,090 miles
    Periapsis: 582,703 kilometers; 362,075 miles
    Semi-Major Axis: 583,520 kilometers; 362,369 miles
    Eccentricity: 0.0014
    Inclination to planet's equator: 0.058 degree
    Period: 13.463234 days
    Average Orbital Speed: 3.15 kilometers/second; 1.96 miles/second

Rotation: synchronous with orbit

Surface Gravity: 0.346 meter/second^2; 0.036g

Spacecraft
    Voyager 2 ....January 26, 1986..............fly-by

Oberon on January 24, 1986 from 660,000 kilometers/410,000 miles.

PARSEC

One parsec is the distance at which the semi-major axis of Earth's orbit subtends an angle of one second.

2.0626481 X 10^8 Astronomical Units
32,313,140,071,200 kilometers
1.9173512 X 10^13 miles
3.2615638 Lightyears

Phobos

(Moon of Mars)

Albedo: 0.071

Apparent Magnitude: +11.4 at planet's mean opposition

Density: 1.876 (water = 1)

Diameter: 27 by 22 by 18 kilometers; 16.8 by 13.67 by 11.2 miles

Escape Velocity: 11.39 meters/second; 37.39 feet/second

Orbit
    Apoapsis: 9517.58 kilometers; 5914 miles
    Periapsis: 9234.42 kilometers; 5738 miles
    semi-major axis: 9376 kilometers; 5826 miles; 2.76 Mars radii
    eccentricity: 0.0151
    inclination to Mars' equator: 1.093 degrees
    Sidereal period: 7 hours 39 minutes 11 seconds, or 0.31891023 day. Synodic period is 13 seconds longer.
    Average Orbital Speed: 2.138 kilometers/second; 1.329 miles/second

Rotation: Synchronous with orbit

Spacecraft

Mariner 7	1969	from space
Viking 1	1977	from surface of Mars
Orbital Global Surveyor	1998	from space
Mars Express	2004	from space
Spirit	2005	from surface of Mars
Mars Reconnaissance	2007	from space

Surface Gravity: 0.005814g

Astronomical Numbers 63

Phobos from 5300 kilometers/3800 miles (roughly the distance from the surface of Mars to Phobos)

Pluto

(dwarf planet; five moons)

Albedo: 0.5 to 0.7 (varied surface)

Apparent Magnitude: +13.65 to +16.3 at opposition

Atmosphere: N_2; CH_4; CO (all normally frozen when planet is far from perihelion)

Density: 1.87 (water = 1)

Discovery: February 18, 1930 by Clyde Tombaugh

Escape Velocity: 1.212 kilometers/second; 0.753 mile/second

Mass: 1.305×10^{22} kilograms

Orbit
    Aphelion: 7,377,590,680 kilometers; 4,584,222,320 miles; 49.31614768 Astronomical Units
    Perihelion: 4,424,306,200 kilometers; 2,749,074,280 miles; 29.573991 Astronomical Units
    Semi-Major Axis: 5,900,898,440 kilometers; 3,666,648,300 miles; 39.4450697 Astronomical Units
    Eccentricity: 0.24897
    Inclination to ecliptic: 17.1405 degrees
    Period: 247.94 years; 90,570 days; 14,164.4 Pluto days
    Average Orbital Speed: 4.67 kilometers/second; 2.902 miles/second

Rotation: 6.38723 days; 6 days 9 hours 17 minutes 36 seconds

Surface Gravity: 0.62 meter/second^2; 0.063g

Spacecraft
    New Horizons .... July 14, 2015 .... fly-by

Pluto on July 14, 2015.

Prometheus
(Saturn XVI; fifth moon outward from Saturn)

Albedo: 0.06

Apparent Magnitude: +15.8 at planet's mean opposition

Axial Tilt: 0

Density: 0.48 (water = 1)

Diameter: 136 by 79.4 by 59.4 kilometers; 84.5 by 49 by 39 miles

Discovery: October 23, 1980, Collins from Voyager 1 photos

Escape Velocity: 0.019 kilometers/second; 62 feet/second

Mass: 1.595×10^{17} kilograms

Orbit
    Apoapsis: 139,420 kilometers; 86,590 miles
    Periapsis: 139,340 kilometers; 86,530 miles
    Semi-Major Axis: 139,380 kilometers; 86,555 miles
    Eccentricity: 0.0022
    Inclination: 0.008 degree
    Period: 0.612990038 Earth day

Rotation: synchronous with orbit

Surface Gravity: 0.0025 meters/second^2; 0.00028g

Note--above is merely an average, as the irregular shape leads to surface gravity varying with location.

Visiting Spacecraft
    Voyager 1 ....fly-by
    Voyager 2 ....fly-by
    Cassini..............................orbit planet

Rhea

(Saturn V; twenty-first moon outward from Saturn)

Albedo: 0.949

Apparent Magnitude: +9.6 at planet's mean opposition

Axis Tilt: 0

Density: 1.236 (water = 1)

Diameter: 1532 by 1525.6 by 1524.4 kilometers; 950 by 948 by 947 miles

Discovery: December 23, 1672 by Giuseppe D. Cassini

Escape Velocity: 0.635 kilometer/second; 0.395 mile/second

Mass: 2.306518 X 10^{21} kilograms

Orbit
    Apoapsis: 527,771 kilometers; 327,942 miles
    Periapsis: 526,445 kilometers; 327,118 miles
    Semi-Major Axis: 527,108 kilometers; 327,334 miles
    Eccentricity: 0.0012583
    Inclination to planet's equator: 0.345 degree
    Period: 4.518212 Earth days
    Average Orbital Speed: 8.48 kilometers/second; 5.27 miles/second

Rotation: Synchronous with orbit

Surface Gravity: 0.027g

Spacecraft

Voyager 1	1980	fly-by
Voyager 2	1981	fly-by
Cassini	November 26, 2005	orbit planet

Rhea on February 10, 2015 from 47,000 miles/76,000 kilometers.

Saturn

(sixth planet from Sun, 62 known moons plus rings)

Albedo: 0.47

Apparent Magnitude: -0.24 to +1.47 (varies with distance, and more importantly, tilt of the rings)

Atmosphere: 96% H2; 3% He; 0.4% CH4; 0.01% NH3; 0.01% HD

Axial Tilt: 26.73 degrees

Density: 0.687 (water = 1)

Diameter

Polar: 108,728 kilometers; 67,560 miles; 8.5521 times Earth
  Equatorial: 120,536 kilometers; 74,895 miles; 9.4492 times Earth

Escape Velocity: 35.5 kilometers/second; 22.06 miles/second

Mass: 5.6836 X 10^26 kilograms; 95.159 times Earth's mass

Orbit
  Aphelion: 10.11595804 Astronomical Units; 1,513,325,783 kilometers; 940,337,000 miles
  Perihelion: 9.04807635 Astronomical Units; 1,353,572,956 kilometers; 841,071,000 miles
  Semi-Major axis: 9.5820172 Astronomical Units; 1,433,449,370 kilometers; 890,704,100 miles
  Eccentricity: 0.055723219
  Inclination: 2.45824 degrees
  Period: 29.4571 years; 10,759.22 Earth days; 24,491.07 Saturn days
  Average Speed: 9.69 kilometers/second; 6.02 miles/second

Rings

D	66,900 to 74,510 kilometers; 41,570 to 46,300 miles
C	74,658 to 92,000 kilometers; 46,390 to 57,160 miles
B	92,000 to 117,580 kilometers; 57,170 to 73,060 miles
A	122,170 to 136,775 kilometers; 75,910 to 84,990 miles
F	140,180 kilometers; 87,100 miles
Janus	149,000 to 154,000 kilometers; 92,580 to 95,700 miles
G	166,060 to 175,000 kilometers; 103,150 to 108,740 miles
Methone	194,230 kilometers; 120,690 miles
Anthe	197,665 kilometers; 122,820 miles
Pallene	211,000 to 213,500 kilometers; 131,000 to 132,660 miles
E	180,000 to 480,000 kilometers; 111,850 to 111,850 to 300,000 mile
Phoebe	4,000,000 to 13,000,000 kilometers; 2,500,000 to 8,000,000 miles (inclination 27 degrees, and retrograde)

Rotation: 10.54 hours; 10 hours 32 minutes 35 seconds

Surface Gravity: 10.44 meters/second^2; 1.065g

Spacecraft

Name	Date	Result
Pioneer 11	September 1979	fly-by
Voyager 1	November 1980	fly-by
Voyager 2	August 1981	fly-by
Cassini/Huyghens	July 1, 2004	orbiter

Saturn from the Casini orbiter, January 2007

SPEED OF LIGHT

The speed of light is traditionally indicated by the letter c (always lower case), from the Latin word celeritas, meaning speed. One of the earliest attempts to measure this speed (which Aristotle had claimed was instantaneous) was performed by Galileo Galilei (1564-1642). He and some assistants set up on the tops of two hills, and flashed lanterns back and forth, while trying to time the flashes. What they discovered was the speed of human response time, thus inventing the start of physiological psychology.

The first true measurement of the speed of light was performed by a Danish astronomer, Ole Roemer, in 1676. He noted that when Earth and Jupiter were on the same side of the Sun (a position known as opposition) eclipses and other events involving the four moons then known around Jupiter came ten minutes earlier than predicted. When Earth and Jupiter were on opposite sides of the Sun (a position known as superior conjunction) such events came ten minutes late.

Roemer deduced that the predicted times were based on Jupiter's average distance from Earth, and that ten minutes represented the time it took light to cross the radius of Earth's orbit.

Speed of light in a vacuum
    299,792.458 kilometers per second
    186,262 miles 2096 feet 5 21/127 inches/second; 186,262.3979 miles/second
    173.144632674 Astronomical Units per day
    1 Astronomical Unit in 8 minutes 19.005 seconds
    1 lightyear per year

Note: Light travels slower in mediums other than a vacuum, by an amount related to the refractive index. Subatomic particles in this situation are sometimes able to move faster than light does, in which case such particles emit Cerenkov radiation.

STARS

Frequency of types among the 100 stars nearest Earth (20 lightyears)
  O and B: 0
  A: 2
  F: 1
  G: 4
  K: 11
  M0 to M2: 12
  M3 to M4: 30
  M5 to M9: 31
  white dwarfs: 6

All are Main Sequence except the white dwarfs. There are no Giant Branch stars within 20 lightyears, although one star, Procyon, appears nearly ready to start evolving onto the Giant Branch.

BRIGHTEST

This list shows star *Name, Constellation, Spectral Class, Apparent Magnitude, Distance* in Lightyears, *Right Ascension,* and *Declination* for the twelve brightest stars in our sky.

Name	Constellation	Class	App Mag	Distance	R.A	Decl.
Sun	-	G2V	-26.7	0.000016	-	-
Sirius	Canis Major	A0V	-1.46	8.6	06h45m08s	-16^42'58"
Canopus	Carina	F01b	-0.72	313	06h23m57s	-52^41'45"
Arcturus	Bootes	K2IIIp	-0.04	37	14h15m40s	+19^10'56"
Acrux	Crux	B0.4IV	+0.04	321	12h26m36s	-63^05'57"
Alpha Centauri	Centaurus	G2V	+0.10	4.36	14h39m36s	-60^50'07"
Rigel	Orion	B8Ia	+0.18	773	05h14m32s	-08^12'06"
Betelgeuse	Orion	M2Ib	+0.3 to +1.2	640	05h55m10s	+07^24'25"
Procyon	Canis Minor	F5IV	+0.38	11	07h39m18s	+05^13'39"
Achernar	Eridanus	B3Vp	+0.50	144	01h37m43s	-57^14'12"
Aldebaran	Taurus	K5III	+0.75 to 0.95	65	04h35m53s	+16^30'34"
Capella	Auriga	G8III	+0.76	42	05h16m42s	+45^59'53"

NEAREST:

The twelve stars nearest Earth giving *Name, Spectral Class, Distance* in Lightyears, *Constellation, Right Ascension,* and *Declination.*

Star	Class	Distance	Constellation	Rt.Ascen	Declin.
Sun	G2V	0.000016			
Proxima Centauri	M5.5Ve	4.243	Centaurus	14h29m42.9s	-62^40'46.1"
Alpha Centauri A	G2V	4.365	Centaurus	14h39m36s	-60^50'02"
Alpha Centauri B	K1V	4.365	Centaurus	14h39m36s	-60^50'02"
Barnard's Star	M4Ve	5.90	Ophiuchus	17h57m49s	+04^41'36"
Wolf 359	M6.5Ve	7.7	Leo	10h56m29s	+07^00'52"
Lalande 21185	M2V	8.24	Ursa Major	11h03m20s	+35^58'12"
Sirius A	A1V	8.58	Canis Major	06h45m09s	-16^42'58"
Sirius B	wd	8.58	Canis Major	06h45m09s	-16^42'58"
Luyten 726-8 A	M5.6Ve	8.73	Cetus	01h19m02s	-17^57'02"
Luyten 726-8 B	M5.5V	8.73	Cetus	01h19m02s	-17^57'02"
Ross 154	M3.5Ve	9.68	Sagittarius	18h49m49s	-23^50'10"

This includes all the stars (but not Brown Dwarfs) within ten lightyears of Earth.

Steins

(asteroid)

Absolute Magnitude: +12.5

Albedo: 0.34

Density: ~1.5

Diameter: 6.67 by 5.81 by 4.47 kilometers; 4.2 by 3.6 by 2.8 miles

Discovery: November 4, 1969 by Nikolai Chernykh

Orbit
    Aphelion: 2.707333 Astronomical Units; 405,011,300 kilometers; 251,662,300 miles
    Perihelion: 2.019845 Astronomical Units; 302,165,500 kilometers; 187,756,300 miles
    Semi-Major Axis: 2.3636 Astronomical Units; 253,589,600 kilometers; 218,710,400 miles
    Eccentricity: 0.145433
    Inclination: 9.9345 degrees
    Period: 1327.26 days; 3.63 years
    Average Orbital Speed: 19.27 kilometers/second; 12.3 miles/second

Rotation: 6.049 hours

Spacecraft
    Rosetta................. September 5, 2008..............fly-by

Sun

Absolute Magnitude: +4.83

Apparent Magnitude: -26.74

Age: 4,350,000,000 years

Chandrasekhar Equations of Energy Production (very simplified)
    Let P stand for Protons. Let N stand for Neutrons.
    Let p stand for positrons. Let e stand for electrons. Let n stand for neutrinos.
    Let G stand for energy. Let the symbol -> stand for "yields". Let . indicate end of process.
    PN is a deuterium atom nucleus. P2N is a tritium atom nucleus. 2P2N is a helium atom nucleus.
    1) P + P -> PN + p + n + G, p + e -> G
    2a) PN + P -> P2N + p + n + G, p + e -> G
    2b) PN + PN -> 2P2N + n + G.
    3) P2N + P -> 2P2N + n + G.

Density: 1.408 (water = 1)

Diameter: 1,392,694 kilometers, 864,500 miles; 108.6 Earths
    Angular Size from Earth: 31.6 to 32.7 arcminutes

Distance from Galactic Center: 27,200 lightyears
    Orbital Period around Galactic Center: ~230,000,000 years
    Orbital Speed: 220 kilometers/second; 130.4 miles/second

Escape Velocity: 617.7 kilometers/second; 388.8 miles/second

Luminosity: 3.846 X 10^{29} Watts

Mass: 1.98855 X 10^{30} kilograms; 330,000 Earths

Obliquity to Ecliptic: 7.25 degrees

Rotation
    Equator: 25.05 days
    Pole: 34.4 days

Spectral Class: G2V

Surface Features
    faculae
    flares
    granulation
    prominences
    sunspots

Surface Gravity: 27.94g

Spacecraft Studying the Sun

Spacecraft	Launch
Pioneer 5	March 11, 1960
Pioneer 6	December 16, 1965
Pioneer 7	August 17, 1966
Pioneer 8	December 13, 1967
Pioneer 9	November 8, 1968
Skylab	May 25, 1973
Helios 1	December 10, 1974
Helios 2	January 15, 1975
Solar Maximum	February 14, 1980
Ulysses	October 6, 1990
Yohkoh	August 30, 1991
Solar & Heliosphere	December 3, 1995
Hinode	September 22, 2006
STEREO 1 & 2	October 26, 2006
Solar Dynamics Obs	February 11, 2010

Helios are German, Yohkoh and Hinode are Japanese. All others are from the United States. There is a small solar telescope aboard the International Space Station.

Tethys

(Saturn III; fifteenth moon outward from Saturn)

Albedo: 0.80

Apparent Magnitude: +10.2 at planet's mean opposition

Axial Tilt: 0

Density: 0.984 (water = 1)

Diameter: 1078 by 1057 by 1052.6 kilometers; 669 by 656 by 654 miles

Discovery: March 21, 1684 by Giovanni D. Cassini

Escape Velocity: 0.394 kilometers/second; 0.245 mile/second

Mass: 6.17449×10^{20} kilograms

Orbit
  Apoapsis: 297,000 kilometers; 184,500 miles
  Periapsis: 291,100 kilometers; 180,900 miles
  Semi-Major Axis: 294,619 kilometers; 182,958 miles
  Eccentricity: 0.0001
  Inclination to planet's equator: 1.12 degrees
  Period: 1.887802 Earth days
  Average Speed: 11.35 kilometers/second; 7.05 miles/second

Rotation: Synchronous with orbit

Surface Gravity: 0.078g

  Spacecraft
  Pioneer 11 ....September 1, 1979 ....fly-by
  Voyager 1 ....November 2, 1980 ....fly-by
  Voyager 2 ....August 26, 1981 ....fly-by
  Cassini................September 24, 2005 ....orbit planet

Tethys from Cassini, July 2007.

Thebe

(fourth moon outward from Jupiter, also Jupiter XIV)

Albedo: 0.047

Apparent Magnitude: +16.0 at Jupiter's mean opposition

Axis Tilt: 0

Density: 0.86 (water = 1)

Diameter: 116 by 98 by 84 kilometers; 71.5 by 61 by 52 miles

Discovery: March 5, 1979 by Stephen Sinnott

Escape Velocity: 25 meters/second; 81 feet/second

Mass: 4.3×10^{17} kilograms

Orbit
    Apoapsis: 226,200 kilometers; 136,900 miles
    Periapsis: 218,000 kilometers; 132,000 miles
    Semi-major axis: 221,889 kilometers; 134,875 miles, 3.11 Jovian radii
    Eccentricity: 0.0175
    Inclination to planet's equator: 1.076 degrees
    Average Orbital Speed: 23.92 kilometers/second; 14.86 miles/second
    Period: 0.674536 Earth Day; 16 hours 11.3 minutes

Rotation: Synchronous with orbit

Surface Gravity: 0.013 meter/second^2; 0.0015g

Spacecraft
    Voyager 1 ....March 5, 1979
    Voyager 2 1979
    Galileo..............................January 4, 2000

Titan

(Saturn VI; twenty-second moon outward from Saturn; Saturn's largest moon)

Albedo: 0.22

Apparent Magnitude: +8.2 at planet's mean opposition

Atmosphere: 98.4% N_2; 1.4% CH_4; 0.2% H_2; trace amounts of C_2H_2

Axis Tilt: 0

Density: 1.8798 (water = 1)

Diameter: 5152 kilometers; 3200 miles; 0.404 Earth

Discovery: March 25, 1655 by Christiaan Huyghens

Escape Velocity: 2.639 kilometers/second; 1.640 miles/second

Mass: 1.3452×10^{23} kilograms

Orbit
    Apoapsis: 1,257,860 kilometers; 789,652 miles
    Periapsis: 1,186,680 kilometer; 736,928 miles
    Semi-Major Axis: 1,221,870 kilometers; 759,218 miles
    Eccentricity: 0.0288
    Inclination to planet's equator: 0.34854 degree
    Period: 15.945 Earth days
    Average orbital speed: 5.57 kilometers/second; 3.46 miles/second

Surface Gravity: 1.352 meters/second^2; 0.14g

Spacecraft
    Cassini..............................orbit planet
    Huyghens ....lander

Titan's Ligeia Mare, the second largest lake on Titan. The liquid is a mix of methane, ethane, liquid nitrogen, and other hydrocarbon liquids. From the Cassini orbiter, April 2007

Titania

(Uranus III; seventeenth moon outward from Uranus; largest moon of Uranus)

Albedo: 0.35

Apparent Magnitude: +13.9

Axis Tilt: 0

Density: 1.711 (water = 1)

Diameter: 1523 kilometers; 946 miles

Discovery: January 11, 1787 by William Herschel

Escape Velocity: 0.773 kilometer/second; 0.48 mile/second

Mass: 3.527×10^{21} kilograms

Orbit
    Apoapsis: 436,390 kilometers; 271,160 miles
    Periapsis: 435,430 kilometers; 270,564 miles
    Semi-Major Axis: 435,910 kilometers; 270,700 miles
    Eccentricity: 0.0011
    Inclination to planet's equator: 0.340 degree
    Period: 8.706234 days
    Average Orbital Speed: 3.646 kilometers/second; 2.266 miles/second

Rotation: synchronous with orbit

Surface Gravity: 0.379 meter/second^2; 0.04g

Spacecraft
    Voyager 2 ....January 26, 1986.............fly-by

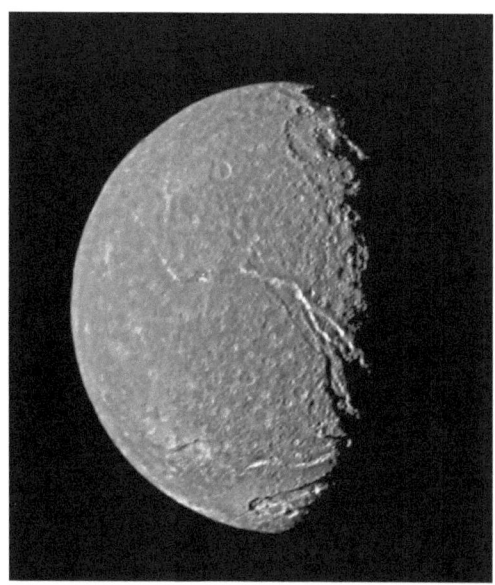

Titania on January 24, 1986 from 369,000 kilometers/229,000 miles.

Triton
(Neptune I; seventh moon outward from Neptune; largest moon of Neptune)

Albedo: 0.76

Apparent Magnitude: +13.47 at planet's mean opposition

Axis Tilt: 0

Density: 2.061 (water = 1)

Diameter: 2707 kilometers; 1680 miles

Discovery: October 10, 1846 by William Lassell

Escape Velocity: 1.455 kilometers/second; 0.904 mile/second

Mass: 2.14 X 10^{22} kilograms

Orbit
    Apoapsis: 354,766 kilometers; 220,440 miles
    Periapsis: 354,753 kilometers; 220,433 miles
    Semi-Major Axis: 354,759 kilometers; 220,437 miles
    Eccentricity: 0.000016
    Inclination to planet's equator: 156.885 degrees (retrograde motion)
    Period: 5.876854 days; 5 days 21 hours 2 minutes 53 seconds

Rotation: Synchronous with orbit

Surface Gravity: 0.779 meter/second^2; 0.0794g

Spacecraft
    Voyager 2 ....1989........................fly-by

Triton from Voyager 2 in 1989 with Neptune in the distant background.

Umbriel

(Uranus II; sixteenth moon outward from Uranus)

Albedo: 0.26

Apparent Magnitude: +14.5 at planet's mean opposition

Axis Tilt: 0

Density: 1.39 (water = 1)

Diameter: 1169.4 kilometers; 726.2 miles

Escape Velocity: 0.52 kilometer/second; 0.32 mile/second

Mass: 1.172×10^{21} kilograms

Orbit
    Apoapsis: 267,037 kilometers; 167,930 miles
    Periapsis: 264,963 kilometers; 164,640 miles
    Semi-Major Axis: 266,000 kilometers; 165,162 miles
    Eccentricity: 0.0039
    Inclination to planet's equator: 0.128 degree
    Period: 4.144177 days
    Average Orbital Speed: 4.67 kilometers/second; 2.90 miles/second

Rotation: synchronous with orbit

Surface Gravity: 0.2 meter/second^2; 0.022g

Spacecraft
    Voyager 2 ....January 1986

Uranus

(seventh planet from Sun; 27 moons plus rings)

Albedo: 0.51

Apparent Magnitude: +5.7 at mean opposition

Atmosphere: 83% H2; 15% He; 2.3% CH4; 0.009% HD; plus ices of NH3, H2O

Axial Tilt: 97.77 degrees

Density: 1.27 (water = 1)

Diameter

Equatorial: 51,118 kilometers; 31,763 miles; 4.007 Earth

Polar: 49,946 kilometers; 31,035 miles; 3.929 Earth

Discovery: March 13, 1781 by William Herschel

Note: Uranus is barely visible to the unaided eye in a dark sky, and it has been claimed the Maori people of New Zealand knew of Uranus before Herschel's discovery.

Escape Velocity: 21.3 kilometers/second; 13.2 miles/second

Mass: 8.681 X 10^25 kilograms

Orbit

Aphelion: 20.095371 Astronomical Units; 3,006,224,700 kilometers; 1,867,981,400 miles
Perihelion: 18,283135 Astronomical Units; 2,735,118,100 kilometers; 1,699,523,600 miles
Semi-Major Axis: 19.189253 Astronomical Units; 2,870,671,400 kilometers; 1,783,752,500 miles
Eccentricity: 0.047220057
Inclination: 0.772556 degree
Period: 84.016846 years; 30,687.15 Earth days; 42,718 Uranian days
Average Orbital Speed: 6.80 kilometers/second; 4.2 miles/second

Rings

Zeta	26,840 to 41,350 kilometers; 16,680 to 25,700 miles
U2R	37,000 to 39,500 kilometers; 22,990 to 24,540 miles
6	41,837 kilometers; 26,000 miles
5	42,234 kilometers; 26,240 miles
4	42,570 kilometers; 26,450 miles
Alpha	44,718 kilometers; 27,790 miles
Beta	45,661 kilometers; 28,370 mile
Eta	47,175 to 47,210 kilometers; 24,310 to 29,335 miles
Gamma	47,627 kilometers; 29,594 miles

Delta	48,295 to 48,306 kilometers; 39,010 to 39,016 miles
Lambda	50,022 to 50,024 kilometers; 31,082 to 31,084 miles
Epsilon	51,100 to 51,200 kilometers; 31,750 to 31,815 miles
Nu	66,100 to 69,400 kilometers; 41,070 to 43,430 miles
Mu	86,000 to 103,000 kilometers; 54,440 to 64,000 miles

Rotation: 0.71833 day; 17 hours 14 minutes 24 seconds

Surface Gravity: 8.69 meters/second^2; 0.886g

Spacecraft

Voyager 2 .... January 24, 1986 ....fly-by

Venus

(second planet from Sun; no moons)

Albedo: 0.67

Apparent Magnitude: -4.9 to black at inferior conjunction

Atmosphere: 96.5% CO_2; 3.47% N_2; 0.015%; SO_2; 0.007% Argon
   Surface Pressure = 92 X Earth

Axis Tilt: 177.36 degrees

Density: 5.243 (water = 1)

Elongation, Maximum: 47 degrees

Escape Velocity: 6.44 miles/second; 10.36 kilometers/second

Mass: 4.8675 X 10^{24} kilograms; 0.815 of Earth

Orbit
   Aphelion: 0.728213 Astronomical Units; 108,939,000 kilometers, 67,691,630 miles
   Perihelion: 0.718440 Astronomical Units; 107,477,000 kilometers, 66,783,170 miles
   Semi-major axis: 0.723332 Astronomical Units; 108,208,000 kilometers, 67,237,910 miles
   Eccentricity: 0.00677323
   Inclination to ecliptic: 3.39458 degrees
   Period: 224.701 Earth days; 0.615198 year; 1.92 Venus days
   Speed, average orbital: 35.02 kilometers/second; 21.76 miles/second
   Synodic period: 583.92 days

Rotation
   243.025 Earth days retrograde
   116.75 Earth days from sunrise to sunrise
   Speed of rotation at equator: 6.52 kilometers/hour; 1.81 meters/second; 4.05 miles/hour

Size
   Diameter: 12,103.6 kilometers; 7520.9 miles; 0.9499 Earth

Surface Gravity: 8.87 meters/second^2; 0.904g

Spacecraft

Name	Date arrived	Effect
Mariner 2	Dec. 14, 1962	fly-by
Venera 3	March 1, 1966	crash on surface
Venera 4	Oct. 18, 1967	entered atmosphere
Venera 5	May 16, 1969	entered atmosphere

Venera 6 ....May 17, 1969 ....entered atmosphere
Mariner 5.....................Oct. 19, 1969.................fly-by
Venera 7 ....Dec. 15, 1970 ....lander
Venera 8 ....July 22, 1972.................lander
Mariner 10..................Feb. 5, 1974 ....fly-by
Venera 9 ....Oct. 22, 1975................lander
Venera 10 ....Oct. 25, 1975................lander
Pioneer Venus.............Dec. 4, 1978.................orbit planet
P.V. Multiprobe ....Dec. 9, 1978..................surface impacts
Venera 11 ....Dec. 25, 1978................fly-by/lander
Venera 12 ....Dec. 19, 1978...............fly-by/lander
Venera 13 ....March 1, 1982..............lander
Venera 14 ....March 5, 1982..............lander
Venera 15 ....Oct. 10, 1983................orbit planet
Venera 16 ....Oct. 11, 1983................orbit planet
Vega 1 ....June 11, 1985 ....atmosphere
Vega 2 ....June 15, 1985 ....atmosphere
Venus Express............April 11, 2006 ....orbit planet
MESSENGER............Oct. 2006.....................fly-by

Mariner, Pioneer Venus, and MESSENGER are from the USA. Venera and Vega were from the USSR. Venus Express was from the European Space Agency.

Venus from orbit showing mainly its northern hemisphere with numerous craters and mountains.

GLOSSARY

Albedo: a measure of the amount of light an object reflects. A perfect mirror would have an albedo of 1.00. Something that reflects no light would have an albedo of 0.00. In general rocks have a low albedo (<.10) while ice and clouds have a high albedo (>.40).

Aphelion: furthest point from the Sun for an elliptical orbit around the Sun

Apoapsis: furthest point from something for an elliptical orbit around that something.

Apogee: furthest point from Earth for an elliptical orbit around Earth.

Apparent Magnitude: a measure of how bright something appears to be. Negative numbers are brighter than positive numbers. The Sun is brightest with an apparent magnitude of -26.7. Generally the unaided human cannot see fainter than apparent magnitude +6.0. Telescopes extend this to as faint as +24.

CH_4: methane

CO_2: carbon dioxide

C_2H_2: acetylene

C_2H_6: ethane

Density: Which would you rather have hit you in the head, a pound of bricks or a pound of feathers? The difference lies in their respective densities.

Eccentricity: A measure of how much an orbit deviates from a perfect circle, which would have an eccentricity = 0.00. Eccentricities equal to or greater than 1.0 mark orbits that escape, i.e. do not remain around whatever their orbit is near.

Escape Velocity: The minimum speed to launch from a surface and boldly go into space, rather than falling back to the ground, or going into orbit around the object just launched from.

Exoplanet: A planet in orbit around any star other than the Sun.

H_2: hydrogen molecule

HD: a hydrogen molecule in which one atom is deuterium (the nucleus contains a neutron as well as the proton).

He: helium atom

Inclination: For planets and asteroids, how much its orbit is tilted with respect to the ecliptic. For moons, how much its orbit is tilted with respect to the planet's equator.

N_2: nitrogen molecule

NaCl: salt

NH3: ammonia

NH4SH: ammonium sulfide

O2: oxygen molecule

PH3: phosgene

Period: time required to complete an orbit

S: sulfur

SO2: sulfur dioxide

Synchronous rotation: The object rotates on its axis in the same amount of time it takes to complete an orbit, thus always having the same face towards whatever it orbits. Mainly found among moons close to their planets.

Credits for photographs: NASA, with the Jet Propulsion Lab of California Institute of Technology

OTHER BOOKS BY THIS AUTHOR

ASTRONOMY

Useful Star Names; with Nebulas and Other Celestial Features. 2011, 70 pages. ISBN 978-61204-614-3

Hundreds of names arranged both by constellation and alphabetically, with the derivation and meaning where known, Right Ascension, Declination, spectral class, distance and apparent magnitudes.

Our Neighbor Stars; including Brown Dwarfs. 2012, 56 pages. ISBN 978-1-61897-132-6

The one hundred stars within twenty lightyears of Earth, their constellations, Right Ascension, Declination, discoverers, sizes, temperatures, known planets, distances, spectral class, and their nearest neighbor. Tables for each of these topics.

Moons of the Solar System; 2013, 70 pages. ISBN 978-1-62516-175-8

185 known moons of the planets and dwarf planets, with names, discovery dates and discoverers, sizes, orbits, orbital periods. Many with photographs made by visiting spacecraft.

Impact Craters of Earth; with selected craters elsewhere. 2014; 57 pages. ISBN 978-1-63135-353-0

About two hundred meteor impact craters are known on every continent of our planet, and even on ocean floors. This is the first book ever to describe all of them, giving location, size, and approximate age. Some are easily visited, while others are unlikely to attract tourists, but this book suggests which may be visited, or avoided. With color photographs of many craters. A few sample craters from elsewhere in the Solar System.

Dwarf Planets and Asteroids; minor bodies of the Solar System. 2014; 69 pages. ISBN 978-1-62857-728-0

The first asteroid was discovered January 1, 1801. Since then hundreds of thousands have been found. This book presents a representative sample, including all asteroids that have been visited by spacecraft, the discoverers, the names, orbits, sizes, and geological make up. About 200 are known to have one or more moons of their own, and these also are covered.

SCIENCE FICTION AND FANTASY

Time for Patriots: Time travel adventure, with appearances by Ben Franklin, George Washington, Wolfgang Amadeus Mozart, and the Bey of Algiers, among many others. 2008, 207 pages. ISBN 978-1-61204-656-3

The Mountain of Long Eyes: 28 stories of science fiction, fantasy, horror, humor and satire. 2012; 213 pages. ISBN 978-1-62212-028-4

Any of these may be ordered directly from the publisher, http://sbpra.com/ThomasWmHamilton

Review Requested:
If you loved this book, would you please provide a review at Amazon.com?

www.ingramcontent.com/pod-product-compliance
Lightning Source LLC
Chambersburg PA
CBHW051156220526
45473CB00003B/795